Unlocking the Power of OPNET Modeler

For fast, easy modeling, this practical guide provides all the essential information you need to know. A wide range of topics is covered, including custom protocols, programming in C++, External Model Access (EMA) modeling, and co-simulation with external systems, giving you the guidance not provided in the OPNET documentation. A set of high-level wrapper APIs is also included to simplify programming custom OPNET models, whether you are a newcomer to OPNET or an experienced user needing to model efficiently. From the basic to the advanced, you'll find topics are easy to follow with theory kept to a minimum, many practical tips and answers to frequently asked questions spread throughout the book, and numerous step-by-step case studies and real-world network scenarios included.

Zheng Lu received his Ph.D. from the University of Essex, after which he stayed on to research optical networks and wireless sensor networks. He is experienced in modeling network protocols and has many years of experience using OPNET Modeler in his research and laboratory demonstrations.

Hongji Yang is currently a Professor at the Software Technology Research Laboratory, De Montfort University. He received his Ph.D. from Durham University in 1994 and was a main contributor to the Distributed Computer Networks project sponsored by the Chinese Ministry of Education, 1982–1986.

Unlocking the Power of OPNET Modeler

ZHENG LU

HONGJI YANG

CAMBRIDGE UNIVERSITY PRESS
Cambridge, New York, Melbourne, Madrid, Cape Town,
Singapore, São Paulo, Delhi, Tokyo, Mexico City

Cambridge University Press
The Edinburgh Building, Cambridge CB2 8RU, UK

Published in the United States of America by Cambridge University Press, New York

www.cambridge.org
Information on this title: www.cambridge.org/9780521198745

First published 2012

Printed in the United Kingdom at the University Press, Cambridge

A catalogue record for this publication is available from the British Library

Library of Congress Cataloguing in Publication data
Lu, Zheng.
  Unlocking the power of OPNET modeler / Zheng Lu, Hongji Yang.
    p. cm.
  Includes bibliographical references and index.
  ISBN 978-0-521-19874-5 (hardback)
  1. Computer networks – Mathematical models. 2. Computer networks – Simulation methods.
  3. Computer network protocols. I. Yang, Hongji. II. Title.
  TK5105.5.L825   2011
  005.7′13–dc23   2011032466

ISBN 978-0-521-19874-5 Hardback

Additional resources for this publication at www.cambridge.org/9780521198745

Contents

Preface

Network simulation is an important methodology in network research fields and OPNET Modeler is a very useful tool for network modeling and simulation. OPNET Modeler is generally used by researchers, protocol designers, university teachers and students in the fields of electronic engineering, computer science, management information systems, and related disciplines. The friendly design of its graphical user interface (GUI) makes it nice and easy to start with. However, the complexity of OPNET Modeler and lack of useful support material make it difficult for many users to fully make use of its benefits. OPNET Modeler has its documentation covering many aspects on using the modeler. However, it covers too many aspects in parallel form rather than a step-forward form, making users unable to decide where to start and causing them to lose focus.

This book is an effort to partially fill this gap and should be useful for courses on network simulation and OPNET modeling for university students, as well as for the researchers on this topic. The book covers a wide range of knowledge from basic topics to advanced topics. All case studies in the book are step-by-step and progressive. Relevant files and sources can be downloaded from the publisher's website. A set of high-level wrapper APIs are provided to help even new users to write complex models, and experienced users to write large, complex models efficiently. Question-and-answer pairs are spread over the chapters to answer the most common questions users may experience in practice.

The book is composed of four parts. Part I: Preparation for OPNET Modeling introduces OPNET and OPNET Modeler. It leads the reader through the required basics on using OPNET Modeler and provides familiarization with OPNET Modeler user interfaces. Part II: Modeling Custom Networks and Protocols first teaches the reader how to create custom models by directly using OPNET API packages. It then introduces a high-level wrapper API package and demonstrates how to model systems easily using these high-level wrapper API packages instead. Part III: Modeling and Modifying Standard Networks and Protocols teaches the reader how to model networks and protocols based on existing standard OPNET modules and how to modify existing standard models in order to extend standard protocols by adding custom features. Part IV: OPNET Modeling Facilities covers content that is used to facilitate OPNET modeling, including debugging, hybrid simulation, External Model Access (EMA), co-simulation, programming OPNET models in C++, etc.

We thank deeply the various people who, during the months over which this endeavor lasted, provided us with useful and helpful assistance. Without their care and consideration, this book would likely not have matured.

First, we thank Dr. David K. Hunter and Dr. Yixuan Qin, who gave us useful suggestions and comments before and during the writing of the book.

Second, we thank the publisher and people who demonstrated interest in publishing this book. The production team at Cambridge University Press has been great. Many thanks go to people who helped us with the book development, including Mrs. Sarah Marsh and Dr. Julie Lancashire.

Dr. Zheng Lu would like to thank his wife Dr. Gui Gui; without her support, he could not have got through that difficult time and thrown himself into finishing the book.

Professor Hongji Yang would like to thank his wife, Xiaodong Zhang, for her full support in finishing the writing of this book.

Trademark acknowledgments: OPNET is a trademark of OPNET Technologies, Inc. All other product names mentioned herein are the trademarks of their respective owners. The relevant screenshots in this book are used with authorization by OPNET Technologies, Inc.

Zheng Lu
Hongji Yang

Abbreviations

API	application programming interface
BSS	basic service set
CDB	Microsoft Console Debugger
CMO	Categorized Memory
DB	diagnostic block
DES	discrete event simulation
EMA	External Model Access
ESA	External Simulation Access
ESD	External System Definition
Esys	External System
ETS	external tool support
FB	function block
FPP	Fractal Point Process
GDB	GNU Project Debugger
GUI	graphic user interface
HB	header block
ICI	Interface Control Information
IDE	Integrated Development Environment
KP	Kernel Procedure
LAN	local area network
MSVC	Microsoft Visual C++ Debugger
ODB	OPNET Simulation Debugger
ODK	OPNET Development Kit
PDF	probability density function
PMO	Pooled Memory
PPP	Point to Point Protocol
QoS	quality of service
RPG	raw packet generator
SDK	software development kit
STD	state transition diagram
STL	Standard Template Library

SV	state variable
TB	termination block
TV	temporary variable
UI	user interface
WAN	wide area network
WLAN	wireless local area network

Part I

Preparation for OPNET Modeling

1 Introduction

This chapter introduces network modeling and simulation, and both OPNET and OPNET Modeler. If you already have relevant background, you can quickly read through this chapter and go to Chapter 2.

1.1 Network modeling and simulation

There are several feasible methods for investigating networking protocols and evaluating network performance (Leemis and Park 2006; www.opnet.com):

- Analysis and mathematical modeling
- Simulation – typically time-based simulation or discrete event-based simulation
- Hybrid simulation with both analysis and simulation
- Test-bed emulation

Analysis and mathematical modeling can provide quick insights and answers to the problems being studied. It is generally faster than simulation, but in many cases is inaccurate or inapplicable. Analytical models are not available for many situations. Even so, many of the available models lack accuracy and some are modeled through approximations. Especially for a network of queues, it can either be decomposed via the Kleinrock independence assumption or be solved using a hop-by-hop single system analysis, both of which lose accuracy. The modeling difficulties and loss of accuracy can be greatly exacerbated when the networking protocols become even slightly complex. It is often necessary to resort to approximation by reducing the general model to a typical and representative analytical path in order to reduce the analytical difficulties (Kleinrock, 1976).

Network simulation provides a way to model the network behaviors by calculating the interactions between modeling devices. Discrete event simulation (DES) is the typical method in large-scale simulation studies instead of a simpler time-based method. DES enables modeling in a more accurate and realistic way, and has broad applicability (Leemis and Park 2006). DES creates an extremely detailed, packet-by-packet model for the activities of network to be predicted. However, it often has significant requirements for computing power; in particular, for very large-scale simulation studies, the process can be time-consuming. It can take several hours or even days to complete. However, simulation can always provide accurate solutions for either a single-node queuing system or a network of queues, from simple algorithm to complex protocol.

One way to work around these issues in mathematical analysis and explicit simulation is combining the methods in the simulation in order to gain access to the advantages of both while overcoming their disadvantages. This combined method is typically called hybrid simulation, i.e., partially modeling in DES for accuracy and partially in mathematical analysis for faster speed and less computational burden (see www.opnet.com).

There are many network simulators like OPNET (see www.opnet.com), NS (www.isi.edu/nsnam/ns), and OMNeT++ (www.omnetpg.org) which are popular and widely used. Among them, OPNET is capable of simulating in both explicit DES and hybrid simulation modes, and supports other simulation features like co-simulation, parallel simulation, high-level architecture, and system-in-the-loop interactive simulations.

Test-bed emulation typically involves implementing the studied algorithms and protocols into real-world hardware but in a much smaller scale or size. Since test-bed emulation considers the aspects of both protocols and real-world situations, it is the best way to provide a benchmark estimating how feasible the algorithms and protocols are and how close they are to the actual situation. Also, it is a useful way demonstrate new networking concepts. The disadvantage is it will also deal with all other real-world difficulties and some unexpected engineering problems which can be completely irrelevant to the studied algorithms and protocols but can be significant in the overall emulation results. Further, the cost of building an emulation test-bed may be significant. Test-beds are not suitable for investigating large systems.

Accordingly, research methodologies for data traffic and networking can be a combination of some or all of these methods. These methods can be used to cross-check each other in order to capture the system in a more accurate, efficient, and cost-effective way.

1.2   Introduction to OPNET

OPNET stands for OPtimized Network Engineering Tools, and was created by OPNET Technologies, Inc., which was founded in 1986. OPNET is a network simulation tool set; its products and solutions address the following aspects of communications networks (see www.opnet.com):

- Application performance management
- Planning
- Engineering
- Operations
- Research and development

This tool set is powerful and can create and test large network environments via software. To address each of these aspects, OPNET provides corresponding product modules throughout its product line.

OPNET products for "application performance management" include ACE Analyst for analytics for networked applications, ACE Live for end-user experience monitoring and real-time network analytics, OPNET Panorama for real-time application monitoring and analytics, and IT Guru Systems Planner for systems capacity management for enterprises.

OPNET products for "network planning, engineering, and operations" include IT/SP Guru Network Planner for network planning and engineering for enterprises and service providers, SP Guru Transport Planner for transport network planning and engineering, NetMapper for automated up-to-date network diagramming, IT/SP Sentinel for network audit, security and policy-compliance for enterprises and service providers, SP Sentinel for network audit, security and policy-compliance for service providers, and OPNET nCompass for providing a unified, graphical visualization of large, heterogeneous production networks for enterprises and service providers.

OPNET products for "network research and development" include OPNET Modeler, OPNET Modeler Wireless Suite, and OPNET Modeler Wireless Suite for Defense.

The products applicable in this book are OPNET Modeler and OPNET Modeler Wireless Suite.

1.3   OPNET Modeler

OPNET Modeler is the foremost commercial product that provides network modeling and simulation software solution among the OPNET product family. It is used widely by researchers, engineers, university students, and the US military. OPNET Modeler is a dynamic discrete event simulator with a user-friendly graphic user interface (GUI), supported by object-oriented and hierarchical modeling, debugging, and analysis. OPNET Modeler is a discrete event simulator that has evolved to support hybrid simulation, analytical simulation, and 32-bit and 64-bit fully parallel simulation, as well as providing many other features. It has grid computing support for distributed simulation. Its System-in-the-Loop interface allows simulation with live systems which feed real-world data and information into the simulation environment. It provides an open interface for integrating external object files, libraries, and other simulators. It incorporates a broad suite of protocols and technologies, and includes a development environment to enable modeling of a very wide range of network types and technologies. With the ongoing release of updated versions, OPNET Modeler incorporates more and more features in order to keep up with the evolution of communication networks, devices, protocols, and applications. Hundreds of protocols and vendor device models with source code are already incorporated in the modeler. OPNET Modeler accelerates the research and development (R&D) process for analyzing and designing communication networks, devices, protocols, and applications (see www.opnet.com). OPNET Modeler GUI makes it user-friendly, and makes it easy for users to begin learning about it and working with it. However, when trying to progress beyond this initial phase, its full-featured functionalities and powerful programming interfaces make it difficult for people to grasp.

OPNET Modeler provides a comprehensive development environment with a full set of tools including model design, simulation, data collection, and data analysis and supporting the modeling of communication networks and distributed systems. OPNET Modeler can be used as a platform to develop models of a wide range of systems. These applications include: standard-based local area network (LAN) and wide area network (WAN) performance modeling, hierarchical internetwork planning, R&D of protocols and communication network architecture, mobile network, sensor network and satellite network. Other applications include resource sizing, outage and failure recovery, and so on.

OPNET Modeler is used in the case studies throughout this book. Readers of this book are assumed to have the license for this particular OPNET product to be able to go through the book content in practice. This book is based on OPNET Modeler 14.5 and later versions, but the modeling methodologies discussed are applicable to earlier versions.

1.4   Summary

This chapter discusses the methodologies for network modeling and simulation. OPNET and its products are introduced. Among OPNET products, OPNET Modeler is the one to address network research and development, and is the product to be used in the case studies throughout this book.

1.5   Theoretical background

1.5.1   Simulation and principles of simulator

Network simulations can be categorized into time-clocked simulation and discrete event simulation. In time-clocked simulation, simulation progresses through the iterative progressing of time slots. Events within the iterated time slot are executed while simulation is progressing. The flowchart of time-clocked simulation is shown in Figure 1.1.

In discrete event simulation, simulation progresses by the execution of the scheduled next event. Simulation time is updated after the next scheduled event is executed. The flowchart of DES simulation is shown in Figure 1.2.

Q1.1   What are the differences between the time slots in time-based simulation and simulation time?

The time slots in time-based simulation refer to the clock time in the real world. Simulation time refers to the time used in running the model. Thus, the simulation time and the time elapsed in a run of the simulation are two different concepts.

Compared with discrete event simulation, time-clocked simulation will iterate all time slots regardless of whether there are events within a particular time slot or not. For a burstlike system with long silent periods, i.e., there are no events in many continuous slots, the time-clocked simulation will be inefficient since it still needs to iterate all those time slots without events being processed. Instead, discrete event simulation only

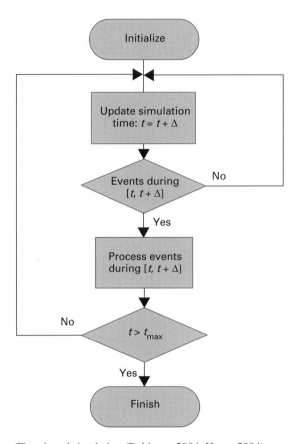

Figure 1.1   Time-based simulation (Robinson 2004, Hayes 2004)

iterates the scheduled events which must be processed in an ordered fashion, to avoid the inefficiency incurred in time-clocked simulation. For this reason, most modern simulators support the approach of discrete event simulation, i.e., DES.

To be able to execute DES, a basic DES simulator framework should have the following elements:

- The random generators representing different random variables as initial system inputs like packet size, packet interarrival times, system processing time and noise, etc.
- Simulation time which can be updated to allow simulation to progress
- Prioritized event lists to store events to be executed one by one
- Simulation finish conditions such as simulation duration, which is the normal way of finishing a simulation, and some other customized termination conditions.

Figure 1.3 shows the pseudo-code of the structure of the basic DES simulator. It has three phases: initialization, simulation, and cleanup. In initialization phase, all state variables are populated with initial values, like simulation time, event list, statistics, and memories. In simulation kernel phase, a main loop is used to run the simulation until a simulation termination condition is reached such as simulation finish time. If

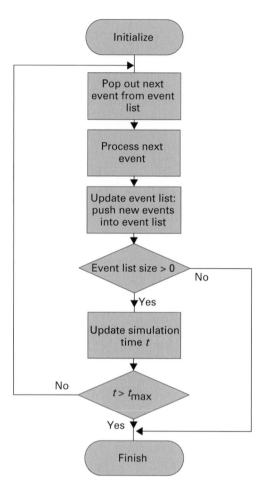

Figure 1.2   Event-based simulation (Robinson 2004, Hayes 2004)

the termination condition is not satisfied, the next scheduled event is popped out and processed. The processing of an event may include: calculating formulas, recording statistics, spawning more events and pushing them into event list, cleaning up invalid variables and free memories, creating new variables and memories, etc. After an event is processed, the simulation time will be updated. The updated value is calculated according to the particular event being executed. This process continues until termination conditions are reached. After leaving the simulation kernel phase, some cleanup work will be done before finishing the simulation, such as writing records into files, freeing memories, and so on.

The simple simulator framework demonstrated in Figure 1.3 is to model a single-process system. However, sometimes systems with concurrent behaviors need to be modeled such as TCP server applications. To model the concurrent behaviors, multiple-processes simulation support will be required inside the simulator's kernel.

```
void main()
{
  //~ initialization
  initialize_variables();
  alocate_memories();
  ...

  //~ simulation kernel operations
  while(simulation_time < finish_time)
  {
    current_event = pop_next_event_from_list();
    process_event(current_event);
    update_simulation_time();
    ...
  }

  //~ finishing up
  write_records_to_file();
  free_memories();
  ...
}
```

Figure 1.3   Pseudo-code for simulation

Q1.2   What are the differences between an operating system process and a simulation process?

An operating system can concurrently run multiple system processes, and a system process can have multiple system threads running within it. A system process must have at least one system thread as its main thread. However, a system thread is lighter than a system process from the perspectives of startup speed, resource occupance, and management burden. Different from a system process, a simulation process here refers to the simulated task such as "transfer of packets," which is an abstract concept used in simulation and is irrelevant to the operating system. In this book, process anywhere refers to a simulation process.

1.5.2   Hybrid simulation

Explicit DES provides accurate simulation results, while analytical methods generally take much less time to compute. Hybrid simulation is a methodology of combining both explicit DES methods and analytical methods in order to take advantages of both. Hybrid simulation may take different forms in actual implementations. In OPNET simulation, the simplest form of hybrid simulation is background traffic simulation. In simulation, traffic through a node can be divided into two parts: explicit traffic and background traffic. Explicit traffic is simulated accurately through the DES method while background traffic is derived analytically. For explicit traffic, each packet's arrival and departure times together with other data of interest are explicitly modeled and recorded. However, for background traffic, there is no tracking of individual packets. The traffic is generated by the workload of the modeled background traffic, and the workload is produced by analytical modeling. Background traffic represents high-level information and is collected over long periods of time. Background traffic is used to characterize and simulate part of a network at an abstracted level in contrast to explicit traffic, which is modeled at a detailed level. The objective in incorporating background traffic is to dramatically reduce the computing power and memory required, in order to save simulation run time.

2 Installation of OPNET Modeler and setting up environments

This chapter shows the steps for installing and configuring the OPNET Modeler and its related environment variables. Having followed this chapter, one should be able to run the OPNET Modeler correctly. If OPNET Modeler and relevant software have already been installed and environment variables have been configured on the target machine, this chapter can be skipped. If you have problems compiling OPNET models, especially, compiling standard OPNET models which should have no compilation and linking errors, please check this chapter to make sure your software is properly installed and environment variables are correctly configured, since many OPNET model compilation and linking errors come from incorrect configuration of the C/C++ compiler's environment variables.

This chapter first describes the system requirements for using OPNET Modeler, including both hardware and software requirements. Then it shows the steps for installing and configuring OPNET Modeler on both Windows and Linux operating systems respectively.

2.1 System requirements for using OPNET Modeler

This section lists the requirements for using OPNET Modeler 14.5 and later versions, and also highlights the relevant key points. For other versions of OPNET Modeler, please check the system requirements datasheet and installation manual shipped with corresponding products, or visit the OPNET website for more information (www.opnet.com). Tables 2.1–2.3 list the system support and hardware and software requirements for using OPNET Modeler.

2.2 Installation on Windows

On Windows, you need to install OPNET Modeler and Microsoft Visual Studio or Microsoft Visual C++ for OPNET Modeler to compile C/C++ based simulation models. The order of installing Visual Studio and OPNET Modeler is irrelevant. However, after finishing installing both of them, you need to check the relevant environment variables to make sure they are correctly configured. If not, you can go through the following steps to make sure everything is installed and configured correctly and ready for modeling and simulation.

Table 2.1 Supported operating systems and processors

Operating system	Processor
Windows 2000 Professional and Server Windows Server 2003 (32-bit and 64-bit) Windows XP Professional (32-bit and 64-bit) Windows Vista Business (32-bit and 64-bit) Red Hat Enterprise Linux 3 and 4 Fedora Linux 3 and 4	x86 or EM64T (Intel Pentium III, 4, Xeon, or compatible), 1.5 GHz or better x86 AMD or AMD64, 1.5 GHz or better

For Windows XP Professional and Windows Vista Business, at least Service Pack 1 is required for OPNET Modeler to work correctly. Since simulation is a computation-intensive process, powerful processors can generally help accelerate the simulation speed.

Table 2.2 General hardware requirements

RAM	Disk space	Display
512 MB is minimum RAM requirement; 1–2 GB RAM recommended	Up to 3–5GB free disk space required for installation	Minimum resolution is 1024×768

For simulating complex models generating large amount of events, more RAM may be required. Running out of memory is quite common in simulation. For disk space, at least several GB free disk space is required for storing simulation files. Sometimes, single simulation scenarios can generate temporary files with several GB. The display resolution is required to allow a simulation graphic user interface (GUI) to be presented in an appropriate form.

Table 2.3 Other requirements

C/C++ compiler	Internet browser	Others
For Linux, gcc 3.4 or higher For Windows, Visual C++ 6.0 or higher	Internet Explorer 5.0 or higher Netscape 7.0 or higher Mozilla Firefox 1.06 or higher	TCP/IP networking protocol support

C/C++ compiler is required to build and debug OPNET models. An internet browser is used for viewing OPNET documentation. The browser should be configured to allow viewing pages with HTML frames and JavaScript. TCP/IP networking software is required to perform some network communication-related tasks like accessing a remote licensing server or serving license for remote clients, and communicating with real network devices. The TCP/IP networking software is generally shipped with the operating system as part of the networking protocol stack.

2.2.1 Installation of OPNET Modeler

The following steps demonstrate how to install OPNET Modeler on a Windows system.

- Log in to Windows as Administrator.

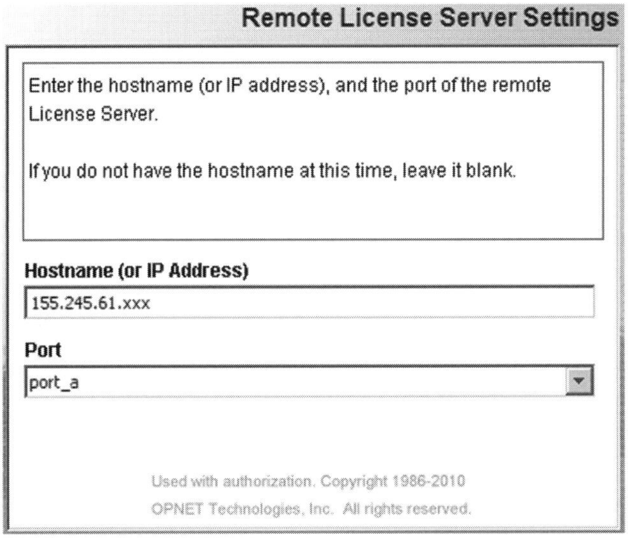

Remote License Server Settings

Enter the hostname (or IP address), and the port of the remote
License Server.

If you do not have the hostname at this time, leave it blank.

Hostname (or IP Address)

155.245.61.xxx

Port

port_a

Figure 2.1   License server settings

- Install OPNET Modeler from the Modeler installation executable file or from the Modeler CD/DVD. During the installation process, you will be prompted to specify the type of licensing scheme:

  - Standalone
  - Floating: access licenses from a remote server
  - Floating: server licenses from this computer.

In "Standalone" mode, one should have license files installed on the local machine and OPNET Modeler can be run only on the machine on which these license files are installed. In "Floating: access licenses from a remote server" mode, the OPNET Modeler installed on the machine will access an OPNET license server either on the local area network (LAN) or on other IP networks. In "Floating: server licenses from this computer" mode, the current machine will act as OPNET license server. For normal OPNET users, one may choose either "Standalone" or "Floating: access licenses from a remote server" scheme. If you choose "Standalone" mode, you can continue installation and you need to add the license or register the product after installation through OPNET Modeler's "License Management" facility. If you choose "Floating: access licenses from a remote server" and press the "Next" button, you will see the dialog as in Figure 2.1. The "Hostname (or IP Address)" refers to the OPNET license server's hostname or its IP address, and "Port" refers to the OPNET license server's predefined listening port. Every time the OPNET Modeler is starting up, it will connect to the specified host and port to acquire licenses. If you do not know your license server, you can just leave them at this moment and specify your license server in OPNET Modeler's "License Management" dialog later after finishing installation.

Figure 2.2   Environment variables

- Install OPNET Models from the models installation executable file or from the Models CD/DVD.
- Install OPNET Modeler Documentation from the documentation installation executable file or from the Documentation CD/DVD.
- Check and configure OPNET environment variables. Open "Control Panel – System dialog" (or right click on "My Computer" and choose "Properties"); select the "Advanced" tab; press the "Environment Variables" button and you will see the dialog as in Figure 2.2.

  In the "System variables" dialog, check whether the environment variables and their values are correctly configured. If any variable is missing, just add it and its value to "System variable" by pressing the "New" button. If the variable is there, but some values are missing, then press the "Edit" button and append these missing values to the end of the "Variable value" string in the pop-up dialog (values are separated by semicolons). Tables 2.4–2.5 list the environment variables that need to be configured for OPNET Modeler. (Note: when you add a value to a variable, you should use a single semicolon to separate them without any space.)

Q2.1   Why should we set environment variables?

The reason for setting these variables is to allow Windows-wide access of relevant files without particularly specifying the paths of these files. For example, if you access the OPNET command "oprunsim" in the command line console without setting the OPNET commands path beforehand, this command cannot be found in command line, as the command line console application will search environment variables for this command. It is the same process for OPNET Modeler to find commands and other dependent files.

Table 2.4 Environment variables for OPNET Modeler

System	Variable	Value
32-bit	PATH	C:\Program Files\OPNET\14.5.A\sys\pc_intel_win32\bin;
64-bit	PATH	C:\Program Files (x86)\OPNET\14.5.A\sys\pc_amd_win64\bin;

The variables set in "User variables" are only applicable for current login user, and not valid for any other users regardless of user account types. The variables set in "System variables" are applicable for all system users. Therefore, to make software function correctly not only for current user but for other users, variables are generally set as "System variables", especially for those software items running as Windows Service.

Table 2.5 Environment variables for Microsoft Visual Studio/Visual C++ 2008 on a 32-bit system

Variable	Value
PATH	C:\Program Files\Microsoft Visual Studio 9.0\Common7\IDE;
	C:\Program Files\Microsoft Visual Studio 9.0\Common7\Tools;
	C:\WINDOWS\Microsoft.NET\Framework\v3.5;
	C:\WINDOWS\Microsoft.NET\Framework\v2.0.50727;
	C:\Program Files\Microsoft Visual Studio 9.0\VC\VCPackages;
	C:\Program Files\Microsoft SDKs\Windows\v6.0A\bin;
INCLUDE	C:\Program Files\Microsoft Visual Studio 9.0\VC\ATLMFC \INCLUDE;
	C:\Program Files\Microsoft Visual Studio 9.0\VC\INCLUDE;
	C:\Program Files\Microsoft SDKs\Windows\v6.0A\include;
LIB	C:\Program Files\Microsoft Visual Studio 9.0\VC\ATLMFC\LIB;
	C:\Program Files\Microsoft Visual Studio 9.0\VC\LIB;
	C:\Program Files\Microsoft SDKs\Windows\v6.0A\lib;
LIBPATH	C:\WINDOWS\Microsoft.NET\Framework\v3.5;
	C:\WINDOWS\Microsoft.NET\Framework\v2.0.50727;
	C:\Program Files\Microsoft Visual Studio 9.0\VC\ATLMFC\LIB;
	C:\Program Files\Microsoft Visual Studio 9.0\VC\LIB;
VSINSTALLDIR	C:\Program Files\Microsoft Visual Studio 9.0;
VS90COMNTOOLS	C:\Program Files\Microsoft Visual Studio 9.0\Common7\Tools;
VCINSTALLDIR	C:\Program Files\Microsoft Visual Studio 9.0\VC;
WindowsSdkDir	C:\Program Files\Microsoft SDKs\Windows\v6.0A;
FrameworkVersion	v2.0.50727
Framework35Version	v3.5
FrameworkDir	C:\WINDOWS\Microsoft.NET\Framework;
DevEnvDir	C:\Program Files\Microsoft Visual Studio 9.0\Common7\IDE

SDK: software development kit

2.2.2   Installation and configuration of Microsoft Visual C++

The following steps demonstrate how to configure environment variables for Microsoft Visual C++ on a Windows system.

- Install Microsoft Visual C++.

Table 2.6 Environment variables for Microsoft Visual Studio/Visual C++ 2008 on a 64-bit system

Variable	Value
PATH	C:\Program Files (x86)\Microsoft Visual Studio 9.0\VC\BIN\amd64;
	C:\WINDOWS\Microsoft.NET\Framework64\v3.5;
	C:\WINDOWS\Microsoft.NET\Framework64\v3.5
	\Microsoft .NET Framework 3.5;
	C:\WINDOWS\Microsoft.NET\Framework64\v2.0.50727;
	C:\Program Files (x86)\Microsoft Visual Studio 9.0\VC\VCPackages;
	C:\Program Files (x86)\Microsoft Visual Studio 9.0\Common7\IDE;
	C:\Program Files (x86)\Microsoft Visual Studio 9.0\Common7\Tools;
	C:\Program Files (x86)\Microsoft Visual Studio 9.0\Common7\Tools
	\bin;
	C:\Program Files\Microsoft SDKs\Windows\v6.0A\bin\x64;
	C:\Program Files\Microsoft SDKs\Windows\v6.0A\bin;
INCLUDE	C:\Program Files (x86)\Microsoft Visual Studio 9.0\VC\ATLMFC
	\INCLUDE;
	C:\Program Files (x86)\Microsoft Visual Studio 9.0\VC
	\INCLUDE;
	C:\Program Files\Microsoft SDKs\Windows\v6.0A\include;
LIB	C:\Program Files (x86)\Microsoft Visual Studio 9.0\VC\ATLMFC
	\LIB
	\amd64;
	C:\Program Files (x86)\Microsoft Visual Studio 9.0\VC\LIB
	\amd64;
	C:\Program Files\Microsoft SDKs\Windows\v6.0A\lib\x64;
LIBPATH	C:\WINDOWS\Microsoft.NET\Framework64\v3.5;
	C:\WINDOWS\Microsoft.NET\Framework64\v2.0.50727;
	C:\Program Files (x86)\Microsoft Visual Studio 9.0\VC\ATLMFC
	\LIB\amd64;
	C:\Program Files (x86)\Microsoft Visual Studio 9.0\VC\LIB\amd64;
VSINSTALLDIR	C:\Program Files (x86)\Microsoft Visual Studio 9.0;
VS90COMNTOOLS	C:\Program Files (x86)\Microsoft Visual Studio 9.0\Common7\Tools;
VCINSTALLDIR	C:\Program Files (x86)\Microsoft Visual Studio 9.0\VC;
WindowsSdkDir	C:\Program Files\Microsoft SDKs\Windows\v6.0A;
FrameworkVersion	v2.0.50727
Framework35Version	v3.5
FrameworkDir	C:\WINDOWS\Microsoft.NET\Framework64;
DevEnvDir	C:\Program Files (x86)\Microsoft Visual Studio 9.0\Common7\IDE;

- Check and configure Visual C++ environment variables to allow OPNET Modeler to function correctly. In "System variables," check whether the variables and their values are correctly configured. Table 2.5 and Table 2.6 list the environment variables to be configured for different versions of Visual C++ under Windows XP. For users under other systems and other versions of Visual C++, check Q2.2 for general solutions. To test whether the Visual C++ compiler is ready for modeling and simulation, after finishing setting up all these variables, you can type "cl" in the

```
@SET VSINSTALLDIR=C:\Program Files\Microsoft Visual...
@SET VCINSTALLDIR=C:\Program Files\Microsoft Visual...
@SET FrameworkDir=C:\WINDOWS\Microsoft.NET\Framework
@SET FrameworkVersion=v1.1.4322
@SET FrameworkSDKDir=C:\Program Files\Microsoft Vis...
@rem Root of Visual Studio common files.
...
@rem
@set DevEnvDir=%VSINSTALLDIR%

@rem
@rem Root of Visual C++ installed files.
@rem
@set MSVCDir=%VCINSTALLDIR%\VC7
...
@set PATH=%DevEnvDir%;%MSVCDir%\BIN;%VCINSTALLDIR%\Comm...
@set INCLUDE=%MSVCDir%\ATLMFC\INCLUDE;%MSVCDir%\INCLUDE...
@set LIB=%MSVCDir%\ATLMFC\LIB;%MSVCDir%\LIB;...

@goto end
...
:end
```

Figure 2.3   Visual Studio environment variables batch processing file

command line console to check whether the compiler is configured correctly, and type "link" to see whether the linker is configured correctly.

Q2.2 How to configure environment variables for other Visual C++ versions in general?

To figure out environment variables for other Visual Studio versions, you can inspect the "vsvars32.bat" or "vsvars64.bat" Windows batch processing file located in "\Common7\Tools" folder within Visual Studio installation path. Assume you'll set up environment variables for Visual Studio 2003. "vsvars32.bat" is the file for setting up environment variables. Its content looks like that in Figure 2.3. You should check "System variables" according to the lines starting with "@set" in the "vsvars32.bat" file. After setting these variables, they will be applicable to Windows-wide applications. (Note: Running the "vsvars32.bat" batch file itself can only set up Visual Studio environment variables for current command line session, not for Windows-wide applications.)

2.2.3   OPNET Modeler preferences for C/C++ compiler

The following steps demonstrate how to configure OPNET Modeler's preferences to support the right C/C++ compiler:

- Start OPNET Modeler and go to "Edit – Preferences" to open the Preferences dialog as shown in Figure 2.4.

Table 2.7 Visual Studio Compilation configurations

Name	Tag	Value
C Compilation Script	comp_prog	comp_msvc
C++ Compilation Script	comp_prog_cpp	comp_msvc
Compilation Flags	comp_flags_devel	/Z7 /Od

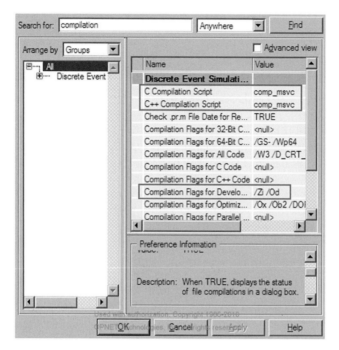

Figure 2.4   Visual Studio environment variables batch processing file

- In Preferences dialog, search for "compilation." Make sure the highlighted attributes and values at the right side of the dialog match the attributes and values in Table 2.7. The first two attributes shown in Table 2.7 are used to set C/C++ compilation script to "comp_msvc.c" C program file which is used to interface OPNET Modeler to Visual C++ compiler. The third attribute in Table 2.7 is used to disable C/C++ code optimization and add debugging information into compiled .obj object files. The third attribute is important when you want to debug your models. By disabling optimization and incorporating debugging information into object files, you are able to track your model code while running your models in debug mode.
- In Preferences dialog, search for the attributes shown in Table 2.8. Make sure their values in Preferences dialog match those in Table 2.8. The first attribute in Table 2.8 is to set C/C++ dynamical linking script to the "bind_so_msvc.c" C program file which is used to interface OPNET to Visual C++ linker in order to generate shared libraries. These shared libraries are ".dll" files on Windows and ".so" files on Linux. The second

Table 2.8 Visual Studio Linking configurations

Name	Tag	Value
Network Repositories Linking Script	bind_shobj_prog	bind_so_msvc
Static Simulation Linking Script	bind_static_prog	bind_msvc
Network Repositories Flags	bind_shobj_flags_devel	/DEBUG
Static Simulation Flags	bind_static_flags_devel	/DEBUG

attribute is used to set C/C++ static linking script to the "bind_msvc.c" file to interface OPNET to Visual C++ linker in static linking mode. The last two attributes are used to set linking options to debug mode, so that debugging information will be added to linked shared libraries for dynamical linking or to executable files for static linking.

Q2.3 Why are there "msvc_binder_error" error, "mspdb??.dll not found" error, "???.h file cannot be found" error, or other similar errors when simulations are running?

It is probably that the environment variables of Visual Studio are not correctly configured. You can check Section 2.2.2 to fix this. If similar problems still exist after configuring the variables, you can check whether you have more than one version of Visual Studio installed on your machine. If so, the variable value of the newer version should be put at the beginning of the value string. This is because OPNET Modeler will look for the first matched variable value. For example, if the first occurrence of compiler path value in "PATH" variable is for Visual Studio 2003 C/C++ compiler, and the first library path value in "LIBPATH" variable is for Visual Studio 2008 library path, then OPNET Modeler will not be able to resolve which version of Visual Studio to use in a consistent manner, i.e., it will use Visual Studio 2003 when it looks for "PATH" variable, but when it looks for "LIBPATH" it expects Visual Studio 2003 library path value to come first rather than Visual Studio 2008 library path.

2.2.4 Licensing

The following steps demonstrate how to make licensing configurations.

- In Section 2.2.1, if the "Floating: access licenses from a remote server" mode is chosen and the right license server and port are set, you can ignore this section. If you do not set the correct license server and port in Section 2.2.1, you can specify the license server and port in OPNET Modeler's "Preferences" dialog as shown in Figure 2.5.
- In Section 2.2.1, if the "Standalone" mode or "Floating: server licenses from this computer" mode is chosen, you need to register your license. Run the "License Manager" program, select "Tools – Register New License" from its main menu, then follow the wizard to complete the license registration process.

If you changed or added environment variables in the above steps, you need to restart OPNET Modeler in order to allow new variables to be refreshed and reflected in OPNET Modeler. Then OPNET Modeler is ready to run simulation models.

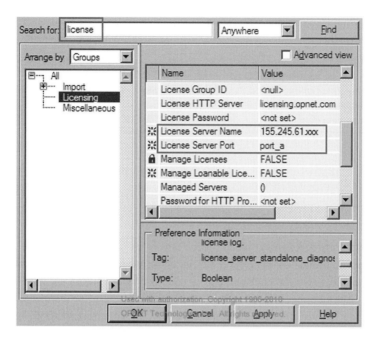

Figure 2.5   Licensing preferences

2.3   Installation on Linux

On Linux, you need to install OPNET Modeler and GCC compiler for OPNET Modeler to compile C/C++ based simulation models. The order of installing GCC compiler and OPNET Modeler is not important. However, after finishing installing both of them, you need to check the relevant environment variables to make sure they are correctly configured. You can go through the following steps to make sure everything is installed and configured correctly and ready for modeling and simulation.

2.3.1   Installation of OPNET Modeler

The following steps demonstrate how to install OPNET Modeler on a Linux system.

- Log in Linux as root.
- Install OPNET Modeler from the modeler installation bin file or from the Modeler CD/DVD. During the installation process, you will be prompted to specify the type of licensing schemes. The specification is the same as in Section 2.2.1.
- Install OPNET Models from the Models installation bin file or from the Models CD/DVD.
- Install OPNET Modeler Documentation from the documentation installation bin file or from the Documentation CD/DVD.
- Check and configure OPNET environment variables. You need to check whether the bin path for OPNET Modeler's executable files is in "PATH" environment variable.

To do so, open a command line shell session, type "echo $PATH" to see if the OPNET bin path is there. If not, you need to set it in the user login initialization file so that once you log in, the OPNET bin path will be automatically set for the user. For a different Linux shell, the initialization file is a little different. To check your shell type, open a shell session and type "echo $SHELL." If it is C-shell, go to your user home path, open ".cshrc" file in an editor and add the following line to the file:

"setenv PATH /usr/opnet/14.5.A/sys/unix/bin:$PATH"

Save the ".cshrc" file and in shell type "source /.cshrc" to force the commands in this file to be executed. Next time, when the user logs into the system, the path configuration commands in the initialization file will be automatically executed. If it is a bash-shell, in your user home path find the ".bashrc" file and add the following line to the file:

"export PATH=/usr/opnet/14.5.A/sys/unix/bin:$PATH"

Save the ".bashrc" file and in shell type "source /.bashrc" to refresh changes.

Q2.4 What are ".cshrc" and ".bashrc" files?

Both files are used to initialize user preferences when the user logs into the system, like setting environment variables, alias, and other setting-up commands. ".cshrc" is for C-shell and ".bashrc" is for bash-shell. These files have "hidden" attribute by default and are located in the user's home directory, i.e., "/[username]/.cshrc" and "/[username]/.bashrc," where "[username]" refers to current login user. The commands in these files will be automatically executed once the user logs into the system.

2.3.2   Installation and configuration of GCC compiler

The following steps demonstrate how to configure GCC compiler on Linux:

- Install GCC compiler if it is not installed. You can check if it is installed by typing "gcc" in shell.
- Check GCC to allow OPNET Modeler to function correctly. Open a shell session, type "gcc -v" to check the GCC version which should satisfy the requirement in Table 2.3. Type "echo $PATH" to see if gcc bin path is there. If not, add it to "$PATH" variable.

2.3.3   OPNET Modeler preferences for GCC compiler

The following steps demonstrate how to configure OPNET Modeler's preferences to support GCC compiler:

- In Preferences dialog, search for the keyword "compilation". Make sure the highlighted attributes and values at the right side of the dialog match the attributes and values in Table 2.9. The first two attributes in Table 2.9 are used to set C and C++ compilation scripts to "comp_gcc.c" and "comp_g++.c" C program files respectively. These two C program files are used to interface OPNET Modeler to the gcc compiler. The third attribute in Table 2.9 is used to produce debugging information for compiled

Table 2.9 gcc Compilation configurations

Name	Tag	Value
C Compilation Script	comp_prog	comp_gcc
C++ Compilation Script	comp_prog_cpp	comp_g++
Compilation Flags	comp_flags_devel	-g

Table 2.10 gcc Linking configurations

Name	Tag	Value
Network Repositories Linking Script	bind_shobj_prog	bind_so_gcc
Static Simulation Linking Script	bind_static_prog	bind_gcc
Network Repositories Flags	bind_shobj_flags_devel	-g
Static Simulation Flags	bind_static_flags_devel	-g

object files. The third attribute is important when you want to debug your models. By setting up these attributes, you are able to track the codes of your models while running your models in debug mode.

- In Preferences dialog, search for the attributes in Table 2.8. Make sure their values in Preferences dialog match those in Table 2.10. The first attribute in Table 2.10 is to set C/C++ dynamical linking script to "bind_so_gcc.c" C program file which is used to interface OPNET to GCC linker to generate shared objects. Shared objects are ".so" binary files on Linux. The second attribute in Table 2.10 is to set C/C++ static linking script to "bind_gcc.c" file to interface OPNET to the gcc linker in static linking mode. The last two attributes in Table 2.10 are used to set linking options to debug mode, so that debugging information will be added to linked ".so" files for dynamical linking or to executable files for static linking.

2.3.4　Licensing

To make licensing configurations on Linux, you should first check if the license directory "/opt/OPNET_license" exists. If not, you need to create that directory. Then you should set the license directory with "read, write and execute" privileges by typing the following command in shell:

> *> chmod 777 /opt/OPNET_license*

Now you can start OPNET license manager by typing the command:

> *> op_license_manager*

Finally, add the license in OPNET License Manager by following the same process as in Section 2.2.4. If you changed or added environment variables in the above steps, you need to restart OPNET Modeler in order to allow new variables to be refreshed and reflected in OPNET Modeler. Then you should be able to start OPNET Modeler to run simulation.

2.4 Theoretical background

2.4.1 Compilation and linking options

Compiler is used to compile source codes into object codes and link object codes into executable, so that the operating system can load that executable into memory to run it. Compiler's compilation and linking options are used to control compiling and linking processes in customized ways. For example, you can change some options to allow codes to be built with debugging information. In OPNET Modeler, the "Development" simulation kernel preference means building simulation models in debug mode, and "Optimized" simulation kernel preference means building simulation models in release mode, i.e., the models are optimized and built without debugging information. With "Development" simulation kernel preference, the simulation process can be debugged but will run slower than "Optimized" simulation kernel. With "Optimized" simulation preference, the simulation will run faster than "Development" simulation kernel, but it cannot be correctly debugged. In OPNET Modeler, you can choose the C/C++ compiler you would like to compile OPNET models with customized compilation and linking options. For more detailed documentation on compiler options, please check the official manuals of the corresponding compiler.

2.4.2 Simulation models compilation and linking

In OPNET simulation, the simulator will follow the same process as normal compilation and linking processes, i.e., the simulator will invoke chosen C/C++ compiler to compile the model code into object code with pre-configured compilation and linking options, then load the compiled models into process to run simulation. Assume "kernel.cpp" includes the implementation of simulator engine and relevant facilities. "library1.cpp" and "library2.cpp" contain the required libraries and other dependent models for simulation. Your own models are stored in the "mymodel.cpp" file. To build a "Development" simulation with debugging capability, the simulation files can be built by entering the following command:

For Visual C++:

> *cl /Od /Zi /OUT:sim_dev.exe kernel.cpp library1.cpp library2.cpp mymodel.cpp*

For gcc:

> *g++ -g -o sim_dev kernel.cpp library1.cpp library2.cpp mymodel.cpp*

Then the executable file sim_dev will be generated and can be loaded into process to run the simulation. However, OPNET encapsulates all the underlying details and many other facilities for you within the Modeler. Therefore, with OPNET Modeler, a user does not need to know the underlying details on how to compile and build simulation models.

3 OPNET Modeler user interface

This chapter walks through graphic user interfaces of OPNET Modeler to help familiarize the reader with the modeler's basic operations. If you are already familiar with OPNET Modeler, its user interface and basic operations, you may ignore this chapter.

The user interfaces described in this chapter include: Project Management Dialog, Modeler Preferences Dialog, OPNET Editors, Simulation Results Browser, and OPNET Documentation Browser.

3.1 Project management

OPNET projects can be easily managed in OPNET Modeler. In "File" menu, a user can choose to create a new project, open an existing project, delete a project, or add a model directory, etc. To create and open a custom project within a directory, you can follow these steps:

- Create a directory where you want your OPNET model project files to be saved. For different projects, you may create individual directories.
- From OPNET Modeler, go to "File – Manage Model Files" menu, choose "Add Model Directory" to add the newly created directory. Then you'll be prompted to confirm model directory as shown in Figure 3.1. You can check both "Include all subdirectories" and "Make this the default directory" options. It is noted that "Make this the default directory" option will force files of other new projects to be saved in this directory. Therefore, it is advisable to always select this option for a new project in order to save the files of the new project into a separate directory.
- From OPNET Modeler, go to "File – Manage Model Files" menu, choose "Refresh Model Directories" to update the new model directory just added. This refresh operation will enable OPNET Modeler to load the new model directory and display it in the project dialog.
- From OPNET Modeler, go to "File" menu, choose "New..." to create a new project. You will be prompted by the "Enter Name" dialog to enter the name for the project and the first scenario of the project. To create a new project using the wizard, in "Enter Name" dialog, you can select "Use Startup Wizard when creating new scenarios" option and press "OK" button to proceed with the project wizard. To manually create a new project, in "Enter Name" dialog, you can unselect "Use Startup Wizard when

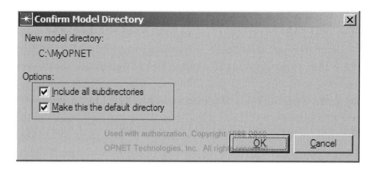

Figure 3.1   Confirm Model Directory dialog

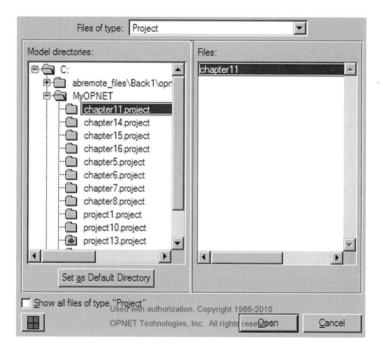

Figure 3.2   Open Project dialog

creating new scenarios" option and press "OK" button to directly go to Project Editor. In Project Editor, you can manually create the project scenario.

- After a new project and scenario are created, in Project Editor, press the "Save" button or enter Ctrl+S to save project files. All project files can be saved into the newly created directory.
- To open an existing project, from OPNET Modeler, go to the "File" menu, choose "Open…" to show an Open Project dialog as in Figure 3.2. From the dialog, you can choose the model directory where your project is located from the left column of the Open Project dialog, then choose the actual project you want to open from the right column of the Open Project dialog.

3.2   Modeler preferences

OPNET Modeler and simulation can be customized via the modeler's Preferences Editor. Preferences Editor allows the user to change the configurations for modeler user interfaces (UI), simulation compiling and linking, trouble shooting, memory management and licensing, etc. From OPNET Modeler, go to the "Edit" menu, choose "Preferences" to open modeler Preferences Editor. Since there are many preferences in the dialog, you may use the "Search for" box to find the preference of interest. In the following are some examples of preferences that a user may be interested to configure at the beginning.

If you want to add/remove/reorder model directories saved in OPNET Modeler, in Preferences Editor you can search for "model directories". Figure 3.3 shows the preference found in Preferences Editor.

You can click the value of this preference to change the model directories as shown in Figure 3.4.

OPNET Modeler 14.5 and later versions support automatic file backup. Therefore, if a user opens an OPNET project and leaves it there, the project files will be automatically saved after some time. This feature is useful to avoid loss of data under some unexpected circumstances such as power outage, program failure, and so on. The default backup interval preference is 60 minutes. You may want to change it to some other value. The preference for changing the backup interval can be found by searching for "backup interval" in Preference Editor as shown in Figure 3.5.

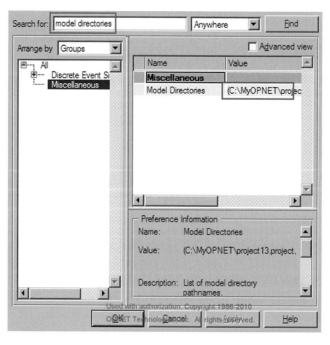

Figure 3.3   Preferences

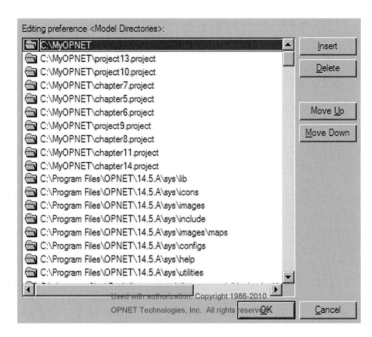

Figure 3.4   Preferences

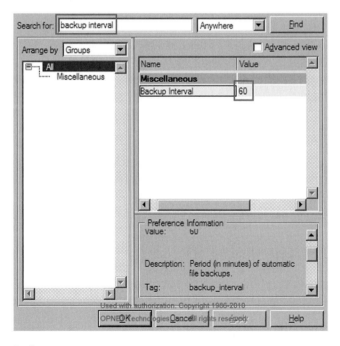

Figure 3.5   Preferences

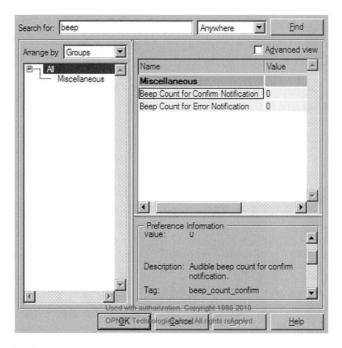

Figure 3.6   Preferences

During simulation, if there are error notifications or confirmation notifications, OPNET Modeler will force your machine to beep. Since this beep comes from the system internal speaker rather than the sound card, adjusting volume will not affect it. If your project is at initial debugging stage, there might be many errors; then every time you run your simulation, it keeps beeping. To stop the beeping, in Preferences Editor you can search for "beep" preference; in the results found, set the preferences "Beep Count for Confirm Notification" and "Beep Count for Error Notification" to 0, as shown in Figure 3.6. Value 0 means "No Beep."

When you edit code in the process model, by default, OPNET Code Editor will be triggered. You can write process model code in this code editor. However, if you are not comfortable with OPNET Code Editor, you can change it via the Preferences dialog as well. In Preferences dialog, you can search for "path to text editor program" preference. Then you can change the value of this preference to the executable path of your favorite code editor. In Figure 3.7, the value is set to the path of Visual Studio program. Then when you open the process model, you can write your OPNET process model code in Visual Studio. If you want to go back to OPNET Code Editor, you can simply change the value of "Path to Text Editor Program" preference to "builtin."

After simulation completes, the user often needs to process the simulation data in a spreadsheet program for analysis. In OPNET Modeler, simulation data can be directly exported to a preselected spreadsheet program. To preselect a spreadsheet program, in Preferences Editor you can search for "path to spreadsheet program" preference and change the value of this preference to the path of your favorite spreadsheet program.

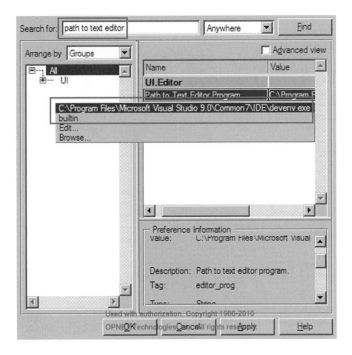

Figure 3.7   Preferences

Then when you export simulation data to the spreadsheet in Simulation Results Browser, the data will be loaded into your specified spreadsheet program. For more information on how to export simulation data into a spreadsheet program, check Section 3.4, "Simulation Results Browser."

There are many other preferences that can customize your modeling and simulation. You can go through all categories of preferences to find the interesting ones.

3.3   OPNET editors

In OPNET Modeler, there are many editors that facilitate and simplify the modeling and simulation tasks by means of easy-to-use graphic user interfaces. In this section, we will go through some of the most frequently used editors including: Project Editor, Node Editor, Process Editor, Link Editor, Packet Format Editor, ICI Editor, PDF Editor, and Probe Editor. For other editors, you can check OPNET documentation for details (www.opnet.com).

3.3.1   Project Editor

Project Editor is the one you may use for every simulation task. Simulation projects and scenarios can be managed by Project Editor. The user can open Project Editor by creating a new project, or opening an existing project. From OPNET Modeler, choose

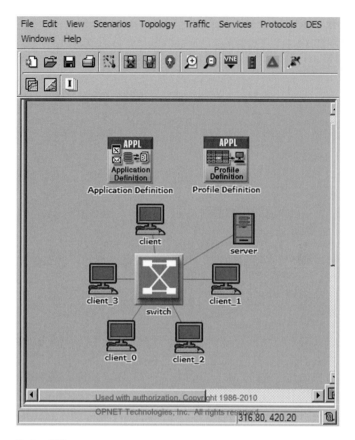

Figure 3.8   Project Editor

"File" menu – "New..." – "Project" to create a new project, or choose "File" menu – "Open..." – "Project" to open an existing project. Figure 3.8 shows the Project Editor with a loaded simulation project scenario.

Project Editor allows you to:

- Create and edit network models
- Create and manage project scenarios
- Configure and import network topology
- Configure and import network traffic
- Customize the network environments
- Verify link connectivity
- Record packet flow animation and node movement animation for subnet
- Configure and run simulations for project scenarios.

These tasks can be achieved by choosing corresponding menu items or clicking the shortcut tool buttons in Project Editor.

OPNET models have a three-layer structure: network model, node model, and process model. Network models are created within Project Editor. You can drag and drop network objects such as subnetworks, network devices/nodes, and links from the Object Palette into Project Editor, and connect them into a network. Node models and process models are created within Node Editor and Process Editor respectively. Node Editor and Process Editor are described in the following sections.

3.3.2 Node Editor

Node Editor is the UI tool by which a user can create and edit the internal structure of a device or node. A node can represent a computer, a switch, a router, or a network cloud. A node is composed of several modules. These modules are generally separated by logical functionalities and are able to communicate with each other via packet streams and statistic streams. Packets can flow through these modules via packet streams. Each module represents a particular functionality of that node. Modules can be used to transmit packets, receive packets, process data, store data, route packets, etc. From OPNET Modeler, you can open an empty node editor by choosing "File" menu – "New..." – "Node Model". You can also open the Node Editor by double-clicking a device or a node in Project Editor. Figure 3.9 shows the Node Editor with a node model loaded.

Node Editor allows you to create and edit modules for the node model. The modules include processor module, queue module, transceiver module, antenna module, and external system module. These modules can be connected by packet streams and statistic streams.

Q3.1 What are the differences between processor module and queue module?

Both processor module and queue module can be used to model the logic process. The only difference is that queue module can also be used to model a buffer, but processor module cannot.

3.3.3 Process Editor

Process Editor is the actual place where you can write code to implement algorithms and protocols. The process model is created and edited in Process Editor. A node model may contain several modules, each of which has a particular functionality. A module should contain a process model that actually implements the functionality or logic this module represents. You can open an empty Process Editor from "File" menu – choose "New..." – choose "Process Model". You can also open the Process Editor by double-clicking a module in Node Editor. Process Editor allows you to visually depict the logic process via state transition diagrams (STDs). In STDs, a logic is composed of several states. States can transition between each other if certain conditions are triggered. You can write C/C++ codes within a state to perform some operations. Figure 3.10 shows an example of STDs of a process model within Process Editor.

Process Editor allows you to implement the actual functionalities by code for corresponding modules.

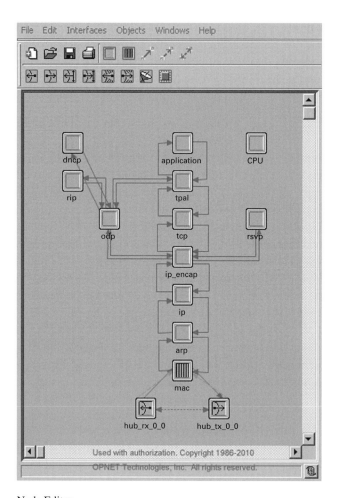

Figure 3.9   Node Editor

3.3.4   Link Editor

Link Editor allows you to create and define a link model. A link model represents a physical connection between nodes. In Link Editor, you can define data rate, bit error rate, channel count, propagation delay, transmission delay, error model, error correction model, supported packet formats, etc. You can open the Link Editor to create a new link model from "File" menu – choose "New..." – choose "Link Model". Figure 3.11 shows the Link Editor with a link model loaded.

Link model supports the following link types: simplex, duplex, bus, or bus tap. It can be configured in "Supported link types" in Link Editor as shown in Figure 3.11.

3.3.5   Packet Format Editor

OPNET Modeler allows you to model packets in both unformatted and formatted forms. For unformatted packets, you can directly create the packet objects in code by invoking

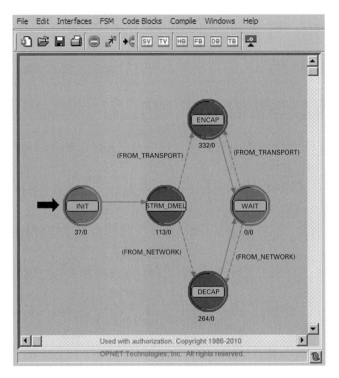

Figure 3.10   Process Editor

unformatted packet APIs. For formatted packets, you should first visually create and format them via Packet Format Editor, then create the formatted packet objects in code by invoking formatted packet APIs. A formatted packet is composed of different fields. In Packet Format Editor, you can specify the type and size of each field. You can open the Packet Format Editor to create a new formatted packet from "File" menu – choose "New..." – choose "Packet Format". Figure 3.12 shows the Packet Format Editor with a packet format loaded.

Fields can be added to Packet Format Editor by using the "Create New Field" button. A user can edit the field's attributes such as type, encoding, size and color, etc. The field type can be: integer, floating point, structure, packet, information, packet ID, or object ID. The supported field types enable the packet to carry any possible information, which may or may not have real-world entity. The encoding can be big-endian or little-endian to model endianness. To edit attributes of a field, right click the field of interest and select "Edit Attributes" from context menu.

Q3.2   Why do we need both formatted and unformatted packets?

For formatted packets, each field in the packet is named. Fields of a formatted packet can be accessed by name. For unformatted packets, each field in the packet is indexed. Fields of an unformatted packet can be accessed by index. Therefore, a formatted packet is often used to model the real-world packet, while an unformatted packet is often used to model a dummy packet or an encapsulated packet.

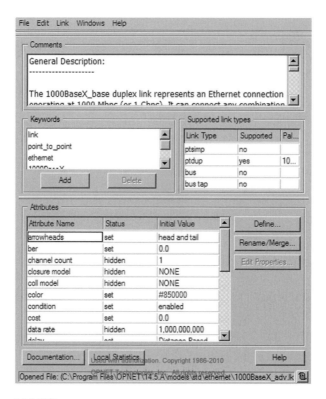

Figure 3.11   Link Editor

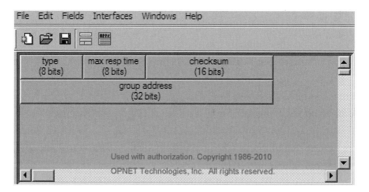

Figure 3.12   Packet Format Editor

3.3.6   ICI Editor

ICI (Interface Control Information) is an OPNET internal structure that is able to carry information and facilitate interrupt-based inter-process communications. A process can access the ICI objects associated with interrupts to communicate with other processes. ICI Editor can be used to visually edit the ICI format. You can open the ICI Editor to

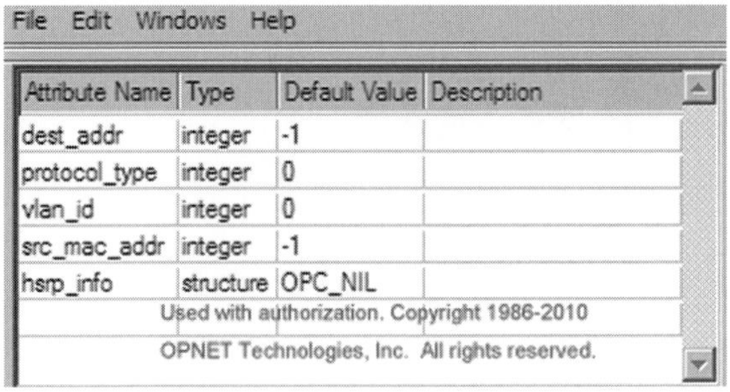

Figure 3.13 ICI Editor

create a new ICI format from "File" menu – choose "New..." – choose "ICI Format."
Figure 3.13 shows the ICI Editor with several attributes defined.

Q3.3 What are OPNET internal structures?

OPNET internal structures are defined for Proto-C. These structures are not general-purpose data structures. They are particularly used in OPNET simulation to facilitate design and development. Other OPNET internal structures include: Packet, Objid, Prohandle, Stathandle, Distribution, and OpT_Packet_Size.

3.3.7 PDF Editor

PDF Editor allows you to create, edit, and view probability density functions (PDFs) of a data sequence. You can load a data sequence into PDF Editor to view and edit. Simulation statistical results can be exported to PDF Editor for analysis as well. You can open the PDF Editor from "File" menu – choose "New..." – choose "PDF Model". Figure 3.14 shows the PDF editor with a data sequence loaded.

You can modify an existing PDF model based on a data sequence, or make a new PDF model. PDF model can be visually edited in PDF Editor by operations such as "Add Impulse", "Normalize", "Set Abscissa and Ordinate Bounds", "Set Sampling Resolution", and "Smooth". The modified or newly created PDF model can be loaded in code by using OPNET Distribution APIs in the process model to model stochastic processes such as system failure, packet size, interarrival times, etc.

3.3.8 Probe Editor

In OPNET simulation, there are many types of statistic – global statistic, node statistic, link statistic, path statistic, etc. However, you may just want to see the ones in which you are interested. To do this, you can use Probe Editor to customize the statistics you want to view after simulation completes. You can open the Probe Editor from "File" menu – choose "New..." – choose "Probe Model". Figure 3.15 shows the Probe Editor loaded with the statistics of interest.

There are different types of statistic you can probe, such as "Global Statistic Probe," "Node Statistic Probe," "Link Statistic Probe," and "Path Statistic Probe." A type of

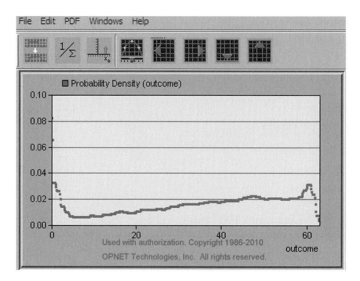

Figure 3.14   PDF Editor

Figure 3.15   Probe Editor

new statistic probe can be created by pressing the corresponding tool button in Probe Editor. For example, if you want to create a statistic probe for a node, you can press the "Create Node Statistic Probe" tool button (the fifth tool button as shown in Figure 3.15). Right click the newly created node statistic probe, and from the Context menu select

"Choose Probed Object" to choose the node objects to probe. The procedure is similar for creating other statistic probes. It is noted that statistics can be collected in different ways. The default one is to collect "All values". This means that during simulation, all statistic points will be recorded. You can choose other, different ways of collecting statistics. To do that, right click the statistic probe of interest, and from the Context menu choose the preferred way of collecting the statistic. For example, if you choose "Collect Time Average Over Default Buckets", the time-averaged statistic will be recorded during simulation instead of the all-values statistic.

3.4   Simulation Results Browser

In OPNET Modeler, for a simulation model, you may have several scenarios to simulate. These scenarios can be based on different topologies, routings, traffic, load parameters, etc. Further, in every scenario you may have many statistics to probe. In OPNET Modeler, simulation Results Browser allows you to view and compare all simulation results for all scenarios of your simulation project in the unified user interface. Simulation Results Browser can be opened from "DES" menu – choose "Results" – choose "View Results...". If your simulation finishes and your model has statistics to probe, you can open Simulation Result Browser to show the statistics in graphs. Simulation Result Browser provides lots of statistical tools to allow you to view and compare the statistical results in different scales such as logarithm, reciprocal, time average, and sample sum. You can also generate distributions from the results, or export results into a spreadsheet for further processing. Figure 3.16 shows the Simulation Results Browser.

To display several statistics in one panel for comparison, you can simply tick the checkboxes of these statistics.

3.5   Animation Viewer

OPNET modeler allows you to record and play animation for packet flows, node movement, and statistic value changes. You can view the recorded animation during the simulation (real-time display), or after simulation finishes. Animation is loaded and played in Animation Viewer. Animation can be controlled in Animation Viewer via operation buttons such as play, pause, stop, restart, speed up, slow down, and skip to next. To open Animation Viewer: in Project Editor, from the "DES" menu choose "Play 2D Animation." If there is animation recorded in this project scenario, Animation Viewer will start and will play the recorded animation automatically. A screenshot of Animation Viewer is shown in Figure 3.17.

You can control the animation using the operation tool buttons in Animation Viewer.

Q3.4   What can one do with Animation Viewer?

Animation Viewer visually demonstrates the process of packet flow and/or node movement. It can help users to visually check if there are some obvious problems in the

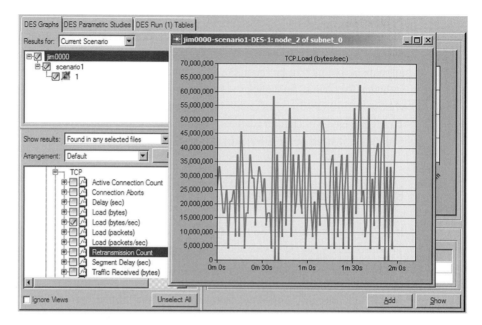

Figure 3.16   Simulation Results Browser

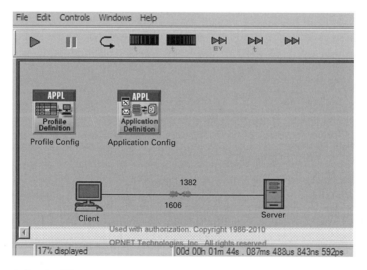

Figure 3.17   Animation Viewer

simulation model. One example: if you model packet routing between several nodes, you can inspect the animation to see if packets are routed in the way you expect. Another example: if you model a mobile network, you can check the animation to see if the mobile nodes move correctly. Animation of the model can also be recorded

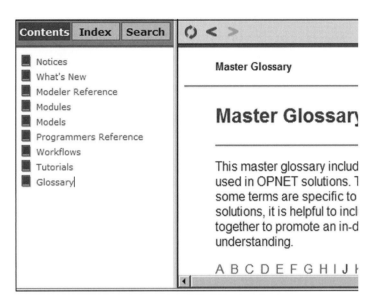

Figure 3.18   OPNET documentation

into video for demonstration using screen recording tools such as SMRecorder (see www.video2down.com) and Snagit (www.techsmith.com/snagit).

3.6   Using OPNET documentation

OPNET documentation provides a very comprehensive description of OPNET Modeler, modeling and simulation. OPNET documentation is managed via a Java-based documentation tool and can be viewed in a standard web browser. The documentation can be opened from "Help" menu of OPNET Modeler. The OPNET documentation tool provides three tools to help you find a particular topic or content. You can find a topic via "Content" tool, "Index" tool, or "Search" tool in OPNET Documentation Browser.

The "Contents" tool lists all topics in a hierarchical structure as shown in Figure 3.18. To check basic knowledge, operations, and user interfaces of OPNET modeler, you can look in the "Modeler Reference" root topic. To check other modeler extension tools, you can look in the "Modules" section. To check OPNET modeling functions/APIs, you can look in the "Programmers Reference" section. For a modeling and simulation tutorial, you can look in the "Tutorials" section. For protocols, algorithms and device models, you can look in the "Models" section.

With the "Index" tool, you can find your topic by following an alphabetical structure, as shown in Figure 3.19.

To quickly find content containing some keywords, you may use the "Search" tool, as shown in Figure 3.20. In the search box, you can enter keywords you want to search for

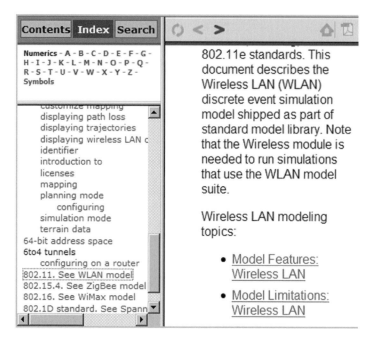

Figure 3.19   OPNET documentation

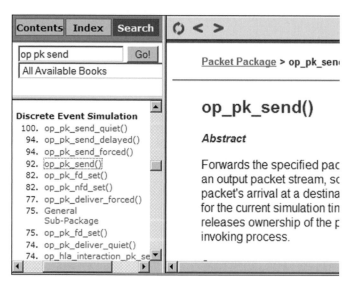

Figure 3.20   OPNET documentation

and hit Return to start searching. There is one thing to note: for some earlier versions of OPNET Modeler, if you want to search some keywords containing special symbols like underline and brackets, you should replace them with a space, otherwise, you will not find any results. For example, if you want to search for "op_pk_send()" function, you should enter "op pk send" instead.

Part II

Modeling Custom Networks and Protocols

4 OPNET programming interfaces

This chapter covers OPNET API (application programming interface) packages and provides an in-depth discussion on a number of commonly used APIs. Practical examples are provided to explain the studied OPNET APIs. If the reader is familiar with OPNET APIs and programming OPNET models, this chapter can be skipped.

4.1 Introduction to OPNET programming

The programming language for writing OPNET Models is called Proto-C. There are not many syntax differences between programming in C and Proto-C, since Proto-C preserves generality by incorporating all the capabilities of the C/C++ programming language, i.e., you can program OPNET models in the same way as you program C/C++ applications. The major difference is the methodology adopted by Proto-C to program models. Unlike programming standalone C/C++ applications, Proto-C is designed to handle OPNET predefined data types via an existing simulation engine, which can be regarded as a half-done application in a standalone C/C++ application. This simulation engine needs to incorporate the Proto-C model code to generate a final runnable and debuggable standalone simulation application. The simulation engine can be regarded as pre-written skeleton or framework which is the kernel in every simulation application. Your OPNET model code is the custom part of the simulation application. Your OPNET model code is inserted into the designated positions of the simulation engine framework to generate the final complete source files. These files will be compiled and linked into a normal C/C++ application. The simulation actually starts only when this application is loaded into the operating system.

However, unlike programming general C/C++ applications, to write OPNET model code you will not directly participate in the whole application-building process, i.e., you do not touch the whole C/C++ sources. What you write is the custom parts of the sources which are abstracted as process models in OPNET Modeler. Further, in the process model, Proto-C adopts a methodology called state transition diagrams (STDs) to help you construct the model logic and handle the underlying C/C++ code. In short, in OPNET programming you do not program the whole sources of an application: just the custom parts of the simulation application. The entire C/C++ simulation application building process will be handled by OPNET Modeler. This service is the most important and kernel service OPNET Modeler provides.

OPNET modeling code can be written in three places:

- The process model via Process Editor or Transceiver Pipeline stages – the OPNET Modeler's kernel GUI services
- Text file via external model access (EMA) interfaces
- Third-party programs via external tool support (ETS) interfaces.

Coding via the process model can handle most tasks. Therefore, we will focus on this in this book. Programming in OPNET can serve different purposes and provide different services. OPNET provides a full set of API packages to serve these purposes. OPNET model developers can use these functions in their code to invoke OPNET simulation services. The breakdown of the major services OPNET provides is as follows.

- Writing discrete event simulation (DES) models like networking algorithms and protocols. This is the kernel service provided by OPNET Modeler and it requires the basic OPNET Modeler license. Writing DES models can take two different forms: (1) via graphic user interface, i.e., the model programs can be written in process model via OPNET modeler's user interface, Process Editor; (2) via text file, external model access (EMA) interface.
- Extending and customizing OPNET Modeler's user interfaces and functionalities, such as adding a new button to OPNET Modeler to perform a user-defined operation. With this capability, users can extend or customize their OPNET Modeler user interfaces to incorporate more features. This capability is known as external tool support (ETS), and now is included in the OPNET Development Kit (ODK) module.
- Others: these include extra flavors to facilitate modeling, design, and presentations such as adding visualization effects, providing co-simulation supports, and writing models into HTML. Some of them may require particular licenses apart from the OPNET Modeler license.

Most OPNET Modeler users who are intending to design and program their algorithms and protocols will generally focus on building discrete-event simulations. To invoke different services, OPNET Modeler provides a couple of categorized API packages.

4.2   OPNET API categorization

This section categorizes various OPNET APIs and explains the differences between these categories to remove confusion about how to choose and use OPNET APIs. OPNET APIs can be categorized into different packages according to their different functionalities: Application Access, Data Structures and Algorithms, Discrete Event Simulation, Generic Runtime System, Model File Access, OPNET Runtime, OPNET Visualization, and Simulation Control. These packages can start with prefixes such as "op_", "prg_", "Ema_", "Ets_", "Oma_", "Osys_", "Ovis_" and "Esa_". Each of these API packages may contain a couple of sub-packages. Some of the packages may provide similar functionalities. However, based on the services they provide, they can be separated into two main groups. One is for writing DES models and providing the kernel services for

OPNET Modeler. The APIs in this group cover different packages such as Data Structures and Algorithms packages, Discrete Event Simulation packages, Generic Runtime System packages, Model File Access packages, OPNET Runtime packages and Simulation Control packages. APIs in this group have prefixes "op_", "prg_", "Ema_" and "Esa_", etc. Another group is for providing extensive services such as extending OPNET user interfaces and functionalities, providing versatile visualization effects and providing extra supports. The APIs in this group include: Application Access packages, Design and Analysis packages, Model File Access packages, and OPNET Visualization packages. In this group, the API prefixes can be "Ets_", "Ace_", "Html_", "Imex_", "Oma_", "Hcon_", "nrac_", "Optim_", "Osys_" and "Ovis_", etc.

Q4.1   How does one find documentation on a particular OPNET API?

To search for a description of OPNET APIs, in some earlier versions of OPNET Documentation you should enter the the API name without underline since the search functionality in these versions cannot interpret non-letter characters. For example, if you wish to search "op_pk_send", you should enter three strings separated by white spaces, i.e., "op pk send" in the search box. For OPNET 16.0 and later, the documentation tool is able to interpret non-letter characters. In this case, you can enter the search term as it is.

Among Group One APIs, the DES packages contain the most important set of APIs, which can be invoked from within process models and provide kernel DES modeling capabilities. These APIs are called Kernel Procedures (KPs) or kernel APIs. OPNET kernel APIs begin with "op_". OPNET kernel APIs perform most OPNET DES Modeling tasks and are capable of providing useful tracing and debugging information.

Q4.2   What licenses are required to write discrete-event simulation (DES) models?

As discussed at the beginning of this section, for different purposes, different API packages and corresponding licenses are required. For programming DES models, only the OPNET Modeler license is required. For other purposes, such as extending OPNET Modeler's user interface, programming visualization and exporting models to HTML, or modeling wireless networks, the individual module licenses are required.

4.3   Kernel APIs/Kernel Procedures (KPs)

In this section, a number of APIs which are difficult to use or are insufficiently documented will be explained and exemplified. The APIs discussed here cover packages of Kernel Procedures, i.e., APIs for writing DES models and providing the kernel services by OPNET Modeler. APIs involved include kernel APIs or Kernel Procedures (KPs) beginning with "op_" and Programming Support APIs beginning with "prg_". To be able to efficiently write DES models, readers should be familiar with these APIs. Topics on Group Two APIs for extending OPNET Modeler user interfaces, i.e., ODK development, are not parts of this book and are not discussed in this section.

The following sub-sections will explain and exemplify the uses of these APIs based on their package categories. Uses of other APIs can be found in OPNET documentation. The exemplification is carried out using complete self-contained code blocks, offering users the ability to reference and copy them directly into their own models. Several APIs may be exemplified in the same code block. The explanations of these APIs within OPNET documentation may involve the Multi-Threading Safety issue, which is an important topic in OPNET parallel simulations. Parallel simulation is used to help accelerate simulation by utilizing multiple CPU processors. The APIs marked as "thread-safe" internally support synchronization of parallel executions, while those marked as "thread-unsafe" must be further processed by users to provide multi-threading synchronization protection or explicitly serialized into a one-after-another manner of execution, under parallel simulation contexts. Multi-threading safety can be ignored by readers if the simulation is not going to run in a parallel mode. In the remaining chapters, if not particularly specified the simulation and its configurations are performed in the single-processor mode, i.e., event executions are performed in a synchronized manner.

As discussed in Section 4.1, there are three main places to write OPNET code. One of them is in the process model or in the Transceiver Pipeline stage. Kernel Procedures should be written in this place. Most simulation model programming tasks can be achieved by writing code in process models and the Transceiver Pipeline stage. In the following chapters, if not explicitly specified the programming-related tasks are programmed in process models and/or Tranceiver Pipeline stages.

The following sub-sections will explain and exemplify the usage of KPs based on their KP package categories.

4.3.1   Distribution Package

Functions within Distribution Package can be used to generate random data with predefined distribution models or with user-defined distribution models. Predefined distribution models can be loaded via the distribution API simply by specifying the distribution model name, as shown in Figure 4.3. To be able to load user-defined or EMA-specified models, it should be first ensured these models are in the Modeler's model directory. If they are in the model directory, they can be loaded via distribution APIs in the same way as loading predefined distributions, i.e., providing the custom model's name without suffix as API input. The APIs do not accept any model arguments for the custom model. This is because the custom model is generated only from experiment samples and is not formulated. Take a custom PDF model with a model file called "mypdf.pd.s" as an example: when this PDF model is loaded via the distribution API, only "mypdf" is to be supplied to the API as parameter. The following paragraphs will describe how to use APIs in this package.

Figures 4.1–4.3 show the loading of predefined, user-defined, and EMA-specified PDF models via distribution APIs.

In Figure 4.1, handles of distribution models are declared. The first three lines are predefined distributions, the fourth line is a user-defined custom distribution, and the last is an EMA-specified distribution.

Figure 4.2 shows the declarations of arguments for predefined distributions.

In Figure 4.3, for predefined distribution, all necessary arguments have been assigned values before invoking "op_dist_load"; for user-defined and EMA-specified

```
/*
 * Declare distribution handles for different PDF models
 * Declared in state variable block
 */
Distribution *dist_uniform;
Distribution *dist_exp;
Distribution *dist_pareto;
Distribution *dist_myuser;
Distribution *dist_myema;
```

Figure 4.1   Code in Process Editor

```
/*
 * Declare and initialize arguments for predefined PDF
 * models
 * Declared in temporary variable block
 */
double uniform_min = 0;
double uniform_max = 0;
double exp_mean = 0;
double pareto_location = 0;
double pareto_shape = 0;
```

Figure 4.2   Code in Process Editor

```
/*
 * Set model arguments and load predefined distributions
 * with arguments.
 * Invoked in states of process model
 */
uniform_min = ...;
uniform_max = ...;
dist_uniform = op_dist_load("uniform", uniform_min, \
    uniform_max);
exp_mean = ...;
dist_exp = op_dist_load ( "exponential", exp_mean, 0.0);
pareto_location = ...;
pareto_shape = ...;
dist_pareto = op_dist_load ( "pareto", pareto_location, \
    pareto_shape);
dist_myuser = op_dist_load("myuser", 0.0, 0.0);
dist_myema = op_dist_load("myema", 0.0, 0.0);
```

Figure 4.3   Code in Process Editor

distributions, no arguments are needed. For a user-defined distribution, its name is "myuser" and its model file should be "myuser.pd.s". For an EMA-defined distribution, its name is "myema" and its model file should be "myema.em.c" correspondingly.

```
/*
 * Declare and initialize variables
 * Declared in temporary variable block
 */
double pksize = 0;
Packet *pkptr = OPC_NIL;
int i = 0;
```

Figure 4.4   Code in Process Editor

```
/*
 * Generate and send 1000 unformatted packets with packet
 * sizes following the loaded exponential distribution.
 * Invoked in states of process model
 */
for(i = 0; i < 1000; ++i)
{
  /* Generate a random packet size following the loaded
   * exponential distribution
   */
  pksize = op_dist_outcome(dist_exp);
  /* Create an unformatted packet with pksize length */
  pkptr = op_pk_create (pksize);
  /* Send this packet to output stream 0 */
  op_pk_send(pkptr, 0);
}
```

Figure 4.5   Code in Process Editor

Figures 4.4 and 4.5 show how to produce random values from the loaded distributions and provide an example of utilizing the generated random values. It is shown that 1000 packets are created with random packet sizes following an exponential distribution.

It is noted that in Distribution Package, there are some APIs whose function names include predefined distribution model names, such as op_dist_exponential() and op_dist_uniform(). They are used to generate random values with exponential and uniform distributions directly, without calling "op_dist_load()" beforehand. They are provided for convenience and as alternatives, since these two predefined PDF models are more frequently used.

After finishing using distributions, these distributions and their allocated memories held by the simulation kernel should be released. This is done by invoking "op_dist_unload()" with the desired distribution handle as its argument, as shown in Figure 4.6. This unloading function can be called once the distribution is not needed in order to save memory. However, when the distribution will be used throughout the whole simulation process, invoking this function is not necessary. This is because after finishing simulation, all allocated memories for current simulation will be automatically freed.

```
/*
 * Unload specified distribution by providing the
 * distribution handle
 * Invoked in states or termination block of process
 * model
 */
op_dist_unload(dist_uniform);
op_dist_unload(dist_exp);
op_dist_unload(dist_pareto);
op_dist_unload(dist_myuser);
op_dist_unload(dist_myema);
```

Figure 4.6   Code in Process Editor

```
/*
 * Define custom structure
 */
typedef struct
{
  int int_value;
  double double_value;
  char string[1024];
} sample_struct;

/*
 * Declare temporary variables
 */
Packet *fmt_pktptr;
Packet *payload_pktptr;
sample_struct *s_structptr;
```

Figure 4.7   Code in Process Editor

4.3.2   Packet Package

Functions in this package are used to create, send, receive, and destroy packets, and to set and get content of packets. The packet can be configured to carry varied information and different types of data, from raw data types such as int, double, char to predefined types and structures such as Objid, OpT_Packet_Id, Packet, Ici, Vvec_Vector, and other user-defined structures. By utilizing procedures in this package, the packet is able to carry any information. This information can be used not only for the purpose of representing packet data, but for purposes of carrying routing table, node information, protocol, and algorithm information, and even other simulation-specific information that is not implemented in real-world packets.

In Figure 4.7, a custom structure is defined and temporary variables are declared. Figure 4.8 shows how to create both formatted and unformatted packets, and how to

```
/*
 * Create packets
 */
// Create a formatted packet and set tota size of this
// packet to 1024
fmt_pktptr = op_pk_create_fmt ( "sample_packet" );
op_pk_total_size_set (fmt_pktptr, 1024);

// Set integer and double value to corresponding fields
op_pk_nfd_set (fmt_pktptr, "sample_field_int", 120);
op_pk_nfd_set (fmt_pktptr, "sample_field_double", \
    120.23);

// Dynamically allocate memory for sample_struct and
// assign structure pointer to a field
s_structptr = \
    (sample_struct *) op_prg_mem_alloc ( \
    sizeof (sample_struct));
op_pk_nfd_set (fmt_pktptr, "sample_field_struct", \
    &s_structptr, op_prg_mem_copy_create, \
    op_prg_mem_free, sizeof(sample_struct));

// Create unformatted packet and assign an integer
// value to the first field of the packet
payload_pktptr = op_pk_create (512);
op_pk_fd_set (payload_pktptr, 0, \
    OPC_FIELD_TYPE_INTEGER, 111, 0);

// Set the unformatted packet pointer to corresponding \
// field of the formatted packet
op_pk_nfd_set (fmt_pktptr, "sample_field_packet", \
    payload_pktptr);

// Send the formatted packet to the output port indexed \
// as 0
op_pk_send(fmt_pktptr, 0);
```

Figure 4.8   Code in Process Editor

assign integer, double, structure, and packet types to fields of a packet. These may have corresponding real-world entities or be only for facilitating modeling. Particularly, to assign a structure to a packet, you have to dynamically create this structure and pass the address of its pointer to the op_pk_nfd_set function. Also, you need to pass the addresses of a memory copy function and a memory free function to op_pk_nfd_set so that it knows how to copy the structure to the packet and how to free the memory of the structure when it is not used. The purpose of this design is to make it possible to assign a hierarchical non-flat data structure to a packet, so that all memories of that structure can be correctly copied and freed via custom copy and free callback functions. However, if the data structure is a flat format, i.e., the structure has no references to other dynamical memories, you can typically use OPNET memory copy and free functions: op_prg_mem_copy_create and op_prg_mem_free.

For creating unformatted packets, you can follow the code in Figure 4.8 directly without any extra work. However, for creating formatted packets, you must first visually create a prototype for that formatted packet via Packet Format Editor, then invoke op_pk_create_fmt in code to create the packet. The name passed to op_pk_create_fmt must match the one created in Packet Format Editor.

Q4.3 What are the differences between op_pk_bulk_size_get/set() and op_pk_total_size_get/set()?

op_pk_bulk_size_get/set() is used to get and set a packet's bulk data size. op_pk_total_size_get/set() is to get and set the total packet size in gross. Bulk data size is a property of a packet that is used to model the amount of data that is not explicitly attributed to individual fields. Modifying the bulk data size of a packet proportionally modifies the total size of the packet, which is the sum of the bulk data size and the sizes of the packet's fields.

Q4.4 What are the differences between op_pk_nfd_access() and op_pk_nfd_get()?

Both functions can retrieve the field value from a packet. op_pk_nfd_get() retrieves the field value and removes the field from the packet. op_pk_nfd_access() only retrieves the field value without removing the field from the packet.

4.3.3 Queue Package and Subqueue Package

In OPNET APIs, a queue object may contain one or more subqueues. The functions in Queue Package are used to access the queue as a whole, while functions in Subqueue Package are used to access an individual subqueue within the queue object. Unlike the queue for a general algorithm, the queue and subqueue here contains only OPNET Packet objects. Functions within these two packages are easy to use. The only point to emphasize is that functions in Subqueue Package support abstract subqueue index and abstract packet position index within a subqueue. Unlike a numeric index, the abstract index can be used to access a subqueue or packet within a subqueue in a particular way. For example, to find a subqueue with the maximum number of packets, the user can write code to loop through all subqueues in order to find the one with most packets by using op_subq_stat function on each subqueue. Alternatively, op_subq_index_map(OPC_QSEL_MAX_IN_PKSIZE) can simply be used to find the numeric index of the subqueue with the most packets. OPC_QSEL_MAX_IN_PKSIZE is the abstract index referring to a subqueue with the maximum number of packets.

4.3.4 Statistic Package

The Statistic Package allows you to read/write customized statistics during simulation. After simulation completes, these statistics can be visualized in Result Browser.

```
// Declare statistics handle as state variable
Stathandle sample_stat_handle;

// Register local statistic and return the handle of
// this statistic
sample_stat_handle = op_stat_reg ("sample_statistic", \
    OPC_STAT_INDEX_NONE, OPC_STAT_LOCAL);

// Write data at current simulation time to statistic
op_stat_write (sample_stat_handle, \
    op_pk_total_size_get(pktptr));
```

Figure 4.9   Code in Process Editor

Figure 4.9 shows how to register a local statistic and write data for current simulation time to that statistic. After simulation completes, you can view this statistic from the Results dialog. The statistic in this example is the current packet size in bits. Local statistic is specified by the OPC_STAT_LOCAL macro. You can also register a global statistic by specifying OPC_STAT_GLOBAL. The difference between a local statistic and a global statistic is that the former allows only the current process to access the statistic, while the latter allows all processes to access the statistic.

It is noted that the statistic name passed to op_stat_reg must match the one created in "Local Statistics" dialog, which can be accessed from the current process model "Interfaces" menu – "Local Statistics."

4.3.5   Segmentation and reassembly package

Segmentation and Reassembly Package provides functions to allow user to split original packets data into segments of any size and reassemble these segments back into the original packets. You can use this package to simulate traffic flows or streams in which traffic is transmitted and routed in units of segments instead of packets. A segment can be spread over several packets or can be part of one packet depending on segment size. In Figure 4.10, segmentation and reassembly buffers are declared and created. Figure 4.11 shows how to create a stream from raw packets and send segments of the stream into the network. Figure 4.12 shows how to add segments into the reassembly buffer and recover raw packets from the reassembly buffer. It is noted that if a segment of a stream is lost during transportation, then the raw packet containing this segment cannot be recovered from the reassembly buffer. The code in Figure 4.11 is implemented at the sender and the code in Figure 4.12 is implemented at the receiver.

4.3.6   Topology package

Topology Package provides functions to allow you to traverse objects like links, nodes and modules in a network. You can use functions in this package to obtain the identifiers

```
// Declare segment buffer and reassembly buffer as
// state variables
Sbhandle seg_buf;
Sbhandle rsm_buf;

// Create segment buffer and reassembly buffer
seg_buf = op_sar_buf_create(OPC_SAR_BUF_TYPE_SEGMENT,\
    OPC_SAR_BUF_OPT_DEFAULT);
rsm_buf = op_sar_buf_create( \
    OPC_SAR_BUF_TYPE_REASSEMBLY, \
    OPC_SAR_BUF_OPT_DEFAULT);
```

Figure 4.10 Code in Process Editor

```
// Add packets into segment buffer to form a stream
for(i = 0; i < 1000; ++i)
  op_sar_segbuf_pk_insert(seg_buf, \
      op_pk_create(1024), i);

// Empty stream by sending all segments from segment
// buffer
for(i = 0; i < 1000*1024/SEGMENT_SIZE; ++i)
  op_pk_send(op_sar_srcbuf_seg_remove( \
      seg_buf, SEGMENT_SIZE), 0);
```

Figure 4.11 Code in Process Editor

```
// Insert received segment info reassembly buffer
op_sar_rsmbuf_seg_insert (rsm_buf, op_pk_get(0));

// Get the total number of original packets within
// reassembly buffer
pkt_num = op_sar_rsmbuf_pk_count (rsm_buf);

// Remove and destroy all original packets from
// reassembly buffer
for(int i = 0; i < pkt_num; ++i)
{
  pktptr = op_sar_rsmbuf_pk_remove (rsm_buf);
  op_pk_destroy(pktptr);
}
```

Figure 4.12 Code in Process Editor

of all objects in the network. Figure 4.13 shows how to obtain the object IDs of the nodes connecting to the current module via packet streams. The current module refers to the module where the current process model resides.

In Figure 4.13, the IDs of the nodes are obtained by crawling from the ID of the current module. Users can easily obtain the IDs of other nodes, modules, or links by modifying these codes appropriately.

4.3.7   Programming Support APIs

Programming Support APIs (begin with 'prg_' prefix) provide many facilities, some of which serve similar purposes to certain OPNET kernel APIs. However, they do not provide tracing and debugging information. In OPNET kernel APIs, those functions provide the same capabilities as the counterparts in Programming Support APIs that start with "op_prg_" prefix. Functions in Programming Support package can be invoked by EMA programs, external tool programs and process model code. The Simulation Kernel Programming package functions can be called only by process model code. Therefore, for DES Modeling via OPNET graphic interface services, i.e. via process models, if there is a kernel API alternative to programming support API, kernel API should always be used since it provides extra debugging information. However, for EMA and ETS programs, Programming Support API prg_??? should be used instead.

4.4   Theoretical background

4.4.1   Proto-C specifications

Proto-C is based on generic C programming language. It supports C programming language specifications and, thus, it can be programmed the same way as programming in C. With Proto-C, the user can invoke existing C library functions as well. However, Proto-C has its own predefined data types and interface library functions (APIs) which distinguish it from generic C language. From this point of view, Proto-C can be regarded as a sub-class of generic C language and is designed specifically for programming simulation models. Further, the Proto-C framework also supports programming in C++; therefore, you can use the standard template library (STL) and Boost libraries in your process models. In all, for programming OPNET models, the user can choose from a variety of API functions including: OPNET APIs, functions defined for OPNET standard models (discussed in Chapter 10), custom wrapper API functions (Chapters 6 and 7), and third-party C/C++ functions like C library, STL, Boost, etc.

Q4.5   What are the differences between C and C++ libraries and OPNET APIs?

Since C and C++ are designed for programming for generic purposes, they only provide standard libraries to achieve functionalities like I/O processing, character/memory handling, basic mathematical calculations, and generic algorithms. However, if a program requires specific functionalities other than those provided by C/C++ standard libraries, such as graphic interface, networking capability, or audio/video processing

```
// Declare temporary variables
int strm_num;
int i = 0;
int j = 0;
int node_num;
Objid strm_id;
Objid tx_id;
Objid link_id;
Objid node_id;

// Get the total number of streams associated with
// current module
strm_num = op_topo_assoc_count(op_id_self(), \
    OPC_TOPO_ASSOC_OUT, OPC_OBJTYPE_STRM);

for(i = 0; i < strm_num; ++i)
{
  // Find the ith stream id
  strm_id = op_topo_assoc(op_id_self(), \
      OPC_TOPO_ASSOC_OUT, OPC_OBJTYPE_STRM, i);

  // Find transmitter id from stream id
  tx_id = op_topo_assoc(strm_id, OPC_TOPO_ASSOC_OUT,\
      OPC_OBJTYPE_PTTX, 0);

  // Find the link id from transmitter id
  link_id = op_topo_assoc(tx_id, OPC_TOPO_ASSOC_OUT, \
      OPC_OBJTYPE_LKDUP, 0);

  // Get the total number of nodes connected by the link
  node_num = op_topo_assoc_count(link_id, \
      OPC_TOPO_ASSOC_OUT, OPC_OBJTYPE_NODE_FIX);

  // Get all nodes ids associated with the link
  for(j = 0; j < node_num; ++j)
  {
    node_id = op_topo_assoc(link_id, OPC_TOPO_ASSOC_OUT, \
        OPC_OBJTYPE_NODE_FIX, j);
    ...
  }
}
```

Figure 4.13   Code in Process Editor

functions, then other non-standard and platform-specific libraries such as GUN C library, Win32 API library, or .NET library may be used. OPNET APIs provide a set of libraries used particularly for simulation, i.e., these libraries are used only in OPNET modeling and simulation and cannot be used for other non-simulation purposes such as programming real-world networking protocols. From this point of view, OPNET APIs are simply another set of C libraries.

Although the OPNET standard models and APIs are implemented in C, OPNET Modeler supports programming in C++ as well. Check Chapter 12 for OPNET programming in C++.

4.4.2   Process model and external model access (EMA) program

OPNET code can be developed in process models or in external model access (EMA) programs.

If a user wants to do general DES modeling, then the code should be written in process models. Programming in process models is able to handle most modeling tasks.

External model access is the technique of accessing a model external to the OPNET analysis software, i.e., accessing a model without using the services provided by the graphical editors. In this context, the definition of accessing a model includes creating the model, modifying the model, and accessing data from the model. External model access is supported via a library of C and C++ accessible functions that serve as a programmatic specification and query language. This library is named the External Model Access (EMA) package and can be viewed as an API for creating and extracting data from model files. EMA provides a text-based alternative to accessing OPNET models instead of graphic interfaces. For how to use EMA, see Chapter 14.

4.4.3   OPNET Modeler model programming external interfaces: co-simulation, external tool support (ETS) and OPNET Development Kit (ODK)

OPNET Modeler's model programming external interfaces refer to the programming ability to allow OPNET Modeler to communicate with external systems. There are several different needs for OPNET Modeler to communicate with external devices, tools, or programs, either hardware, software, or both. To suit these different needs, OPNET Modeler provides different external interfaces.

To facilitate co-simulation with other programs, OPNET provides an external system definition (ESD) model. An ESD model defines a set of interfaces that allow process models in OPNET modeler to communicate with external programs. These interfaces can be read or written by both OPNET process models and external programs.

To allow users to extend the interfaces and functionalities of OPNET Modeler, OPNET provides a set of programming packages. The Application Access Package is used with OPNET Development Kit (ODK) to extend OPNET Modeler's default user interface and functionalities, e.g., ODK and Application Access Package can be used to create individual services in OPNET Modeler, like adding a new menu item in the modeler's menu list to perform customized operation. Extending OPNET interface and functionalities are not topics covered in this book. Readers who are interested in these topics can refer to ODK documentation. A separate license is required to be able to use the ODK module. ODK is inherited from ETS, which was originally part of OPNET Modeler, and now ODK replaces and extends ETS and becomes another OPNET module. The ETS package was introduced in OPNET 7.0 to interact with user-defined modeler interface elements. In the subsequent versions of OPNET, ETS evolved into a formal and complete customization environment for extending OPNET Modeler's user interface and functionalities and is named OPNET Development Kit (ODK).

To avoid confusion, note that:

- External model access (EMA) is not OPNET Modeler's external interface; it is defined as a technique for accessing OPNET models via a non-graphic interface rather than accessing OPNET models via Modeler's graphic editors, such as Network Editor, Node Editor, Process Editor and Packet Editor, etc. Therefore, in EMA, "external" refers to text-based model access ability external to OPNET Modeler's graphic model access ability.
- External interfaces here particularly refer to the OPNET programming ability to allow OPNET models to communicate with external systems, tools or programs, which is different from other general external interface concepts like external debugger support.

5 Creating and simulating custom models using OPNET APIs

This chapter shows how to create and program custom models in OPNET Modeler with progressive case studies, to help readers gradually build up their knowledge on custom model creation. The chapter covers basic knowledge and techniques on custom model creation, model optimization, simulation, results visualization, comparison, and analysis. If readers already know the basics of simulation and how to create custom models, this chapter can be skipped.

5.1 General procedure for creating and simulating custom models

There are several ways to create simulation models and execute simulation. However, in this section, a general procedure of doing these via OPNET Modeler's GUI is introduced.

Q5.1 What are OPNET models?

OPNET models include node model, process model, link model, path model, network model, packet format models and ICI format model. Models are saved in .m files. For a node model, the file extension is ".nd.m." For a link model, the file extension is ".lk.m." For a process model, the file extension is ".pr.m." These model files are saved in the model directory, which can be found from OPNET Modeler: choose menu "Edit" – "Preferences" – search for "Model Directories."

First, custom models need to be created, and then a simulation scenario to test the custom models is created.

The following are the steps for creating custom models:

- Design the custom node in Node Editor by separating different logic functionalities into different modules.
- Implement process models of these modules via a state transition diagram (STD) in Process Editor.
- Optimize and validate models.
- Compile and debug process models.

The following are the steps for simulating custom models:

- Create a project in Project Editor.
- Create a scenario within this project.

- Create a network topology for this scenario by placing the created custom models on the Project Editor.
- Verify link connectivity of this network.
- Select statistics of interest.
- Run simulation for this scenario.
- View, compare, and analyze statistic results.
- If necessary, export statistics data to a spreadsheet for further processing.

It is not necessary to follow the above steps exactly in the given order. For model optimization and validation, they are always performed repetitively throughout the modeling life cycle.

5.2   Custom models

In this section, progressive case studies are provided to help readers understand the necessary concepts for creating custom models. Creating custom models generally involves creating custom node models and process models. Before any scenario is prepared, the model directory that can be used to store the project and model files needs to be checked. By default, the project and model files will be saved in pre-defined model directory or standard model directory. To change default model directory, from "Edit" menu, choose "Preferences." In Preferences Editor, search for "Model Directories." In the "Model Directories" dialog, insert the new model directory before the current first directory, because the first directory in the "Model Directories" dialog is the default model-saving directory. Then, all project and models files will be placed in this new model directory.

Q5.2   Why are there write permissions errors when running simulation?
     This is because the relevant model directories have no write permissions. The relevant model directories can be either standard model directory or custom model directory. So check all directories containing models that are used in the simulation scenario, to make sure they all have write permissions.

5.2.1   Case 1

Case 1 creates a custom traffic source node model and relevant process models. To create a new custom node model, from "File" menu, choose "New..." – "Node Model." This will open Node Editor. In the editor, add four modules: two processor modules, a point-to-point receiver module, and a point-to-point transmitter module. Create packet streams (blue lines) between these modules. The packet stream lines allow packets to flow between these modules following stream line direction. Create logic Tx/Rx association (orange line) between receiver and transmitter. The association is used to bind a particular receiver module to a particular transmitter module to make them work in duplex mode. This custom node is shown in Figure 5.1.
     Finally, save this node model as "basic_source."

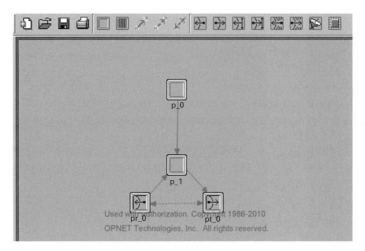

Figure 5.1   Custom node model

Q5.3   Can custom link models be created?

Yes. Individual pipeline models can be created to model customized features of the link. To create pipeline models, from "File" menu, choose "New..." – Pipeline Stage (C code) or (C++ code) to open a new pipeline code editor. Corresponding code to implement customized link features such as link error model and propagation delay model can be created. After the pipeline models have been completed and saved, these pipeline models in a link model can be chosen. To do this, right-click a link model, choose "Edit Attributes (Advanced)"; for the attributes such as "ecc model," "error model," and "propdel model," the corresponding pipeline models just created can be chosen.

Right click processor module "p_0," choose "Edit Attributes" to show attributes table. For "process model" attribute, choose "simple_source." For "Packet Interarrival Time," choose "exponential." Now, module "p_0" is able to generate Poisson traffic with exponential packet interarrival times.

Next, a new process model for the processor module "p_1" can be created so that this module is able to forward packets to the transmitter and destroy packets from the receiver. From "File" menu, choose "New..." – "Process Model" to show Process Editor. Process Editor enables the creation of process models via state transition diagrams (STDs), which separates the process logic into different states and makes these states connected to each other via state transition lines. Add four states to the process editor. Make three states forced, connect these states via transition lines, and set the names for these states, as shown in Figure 5.2.

Q5.4   What are the differences between forced and unforced states?

A state has two parts for code execution: enter part and exit part, which are represented by the upper half and the lower half of the state circle respectively. A forced state will execute the code in both enter part and exit part, then transition to another state via the state transition line. An unforced state will execute only code in the enter part and pause

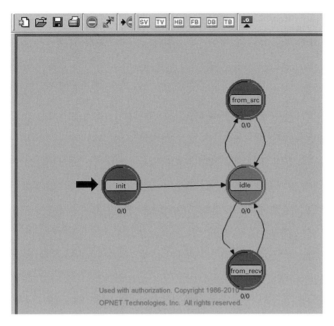

Used with authorization. Copyright 1986-2010
OPNET Technologies, Inc. All rights reserved.

Figure 5.2   Process model

until a further interrupt is triggered by the transition condition. Then it executes the code in the exit part and transitions to the next state. A forced state is in green color and an unforced state is in red. State can be set to forced or unforced by right-clicking state and choosing "Make State Forced" or "Make State Unforced."

The three forced states transition to "idle" state without a transition condition. However, unforced "idle" state needs two transition conditions to allow it to transition to "from_src" and "from_recv" states. In an OPNET process model, the transition condition can be set in the transition line's "condition" attribute and defined as a C macro in header block (HB). For transition from "idle" state to "from_src" state, set the transition condition to "UPPER_STRM." For transition from "idle" state to "from_recv" state, set the transition condition to "LOWER_STRM." For "idle" state, add a default transition to the state itself; this is to avoid possible transitions without true value. The STD is shown in Figure 5.3.

At this point, the state transition diagram of this process model is completed. Save this process model as "traffic_source." This process model will be used in "p_1" module. The logic flow of this state transition diagram is as follows: the control initially enters "init" state and executes the code in "init" state, then transitions to "idle" state; the control waits in "idle" state until either "UPPER_STRM" or "LOWER_STRM" condition is triggered; once the transition condition is triggered, state will transition from "idle" state to either "from_src" or "from_recv" state, and then transition back to "idle" state to wait for new interrupt. This process will repeat itself. It is noted that a process model is not restricted to a particular state transition diagram, i.e., you may implement a process

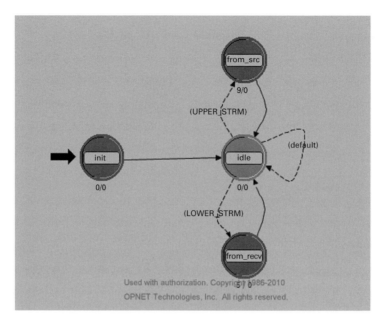

Figure 5.3   Process model

```
#define UPPER_IN_STRM_INDEX 1
#define LOWER_IN_STRM_INDEX 0
#define LOWER_OUT_STRM_INDEX 0
#define UPPER_STRM ( \
    (op_intrpt_type() == OPC_INTRPT_STRM) && \
    (op_intrpt_strm() == UPPER_IN_STRM_INDEX))
#define LOWER_STRM ( \
    (op_intrpt_type() == OPC_INTRPT_STRM) && \
    (op_intrpt_strm() == LOWER_IN_STRM_INDEX))
```

Figure 5.4   HB block

model via different STDs to reflect the same logic. In this case study, only one possible STD for describing a process model is demonstrated.

Next, code is added to this process model. The code can be placed in several places: states (both enter part and exit part), SV, TV, HB, FB, DB, TB code blocks. First, add the macro definitions for the state transition conditions "UPPER_STRM" and "LOWER_STRM", which are actually triggered by process model interrupts. This is shown in Figure 5.4.

Q5.5   What are SV, TV, HB, FB, DB, and TB in Process Editor?

SV, TV, HB, FB, DB, and TB are several code blocks where C/C++ code can be placed. These code blocks can be edited via corresponding toolbar buttons on Process Editor. SV stands for state variable. In OPNET process model, state variable keeps valid between state transitions. TV stands for temporary variable. A temporary variable is only valid within a state. HB stands for header block. A header block is similar to a C/C++ header file which can be used to include other header files and define macros, etc. FB stands for function block, which contains definitions of functions. DB stands for diagnostic block, which contains diagnostic statements that print out diagnostic information to the standard output device. The diagnostic block procedure can be triggered by the OPNET Simulation Debugger (ODB). For more information on debugging in OPNET Modeler, check Chapter 11. TB stands for termination block. The code in this block will be executed just before the current process is destroyed. Therefore, readers can write termination code in this block.

Q5.6   What is interrupt in process model?

Interrupt in process model is used as a way to switch the direction of simulation execution. In process model, code execution can be switched to different states by triggering an interrupt. There are many types of interrupt: packet stream interrupt, statistic interrupt, self-interrupt, etc. Self-interrupt can be used to schedule users' own interrupts at specified simulation times in order to execute the desired code.

In Figure 5.4, both "UPPER_STRM" and "LOWER_STRM" are defined as stream interrupt. The difference is that the input stream index for "UPPER_STRM" is 1 and for "LOWER_STRM" is 0. The stream index can be identified in Node Editor by checking the packet stream's "src stream" or "dest stream" attribute, depending on whether the stream is input or output relative to the studied module; in this case the studied module is "p_1," as shown in Figure 5.5.

In Figure 5.5, it is shown that there are two packet streams flowing into "p_1" module. The packet stream from "p_0" to "p_1" has "dest stream" value 1 and the packet stream from "pr_0" to "p_1" has "dest stream" value 0. These two values correspond to the

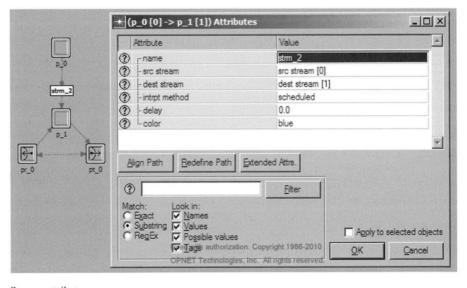

Figure 5.5   Stream attributes

```
pk = op_pk_get(UPPER_IN_STRM_INDEX);
op_pk_send(pk, LOWER_OUT_STRM_INDEX);
```

Figure 5.6   "from_src" state

```
Packet *pk = OPC_NIL;
```

Figure 5.7   TV block

```
pk = op_pk_get(LOWER_IN_STRM_INDEX);
op_pk_destroy(pk);
```

Figure 5.8   "from_recv" state

input stream indices for "UPPER_STRM" and "LOWER_STRM" respectively. There is another packet stream flowing from "p_1" to "pt_0." The "src stream" value is 0, which is the output stream index for the "p_1" module.

In "from_src" state, one can add the code shown in Figure 5.6 to get the packet in the input packet stream from "p_0" module and send the packet to the output stream to "pt_0" module.

In a TV code block, one can add the declaration of the packet temporary variable, as in Figure 5.7.

In "from_recv" state, one can add the code shown in Figure 5.8 to destroy the packet in the input packet stream from "pr_0" module.

From menu "Interfaces," choose "Process Interfaces." Set "begsim intrpt" and "endsim intrpt" to "enabled", so that the process model will be started by a begin simulation interrupt and finished by an end simulation interrupt instead of other interrupts.

By pressing the toolbar button labelled "Compile Process Model" compile the process model into an object file, which will be linked with other simulation object files when simulation is starting. The compilation will generate the C or C++ source file (.pr.c or .pr.cpp) of the process model in the model directory. If there is a compilation error, follow the error message to correct the error and compile it again. Repeat this process until no compilation error is reported.

In "basic_source" node editor, set the "process model" attribute of "p_1" module to "traffic_source" and save "basic_source" node model.

At this point, a custom node model has been created that is capable of generating and forwarding packets to its transmitter and destroying packets received from the receiver. Next, a new link model is going to be created that is able to connect the custom nodes.

From "File" menu, choose "New..." to create a new Link Model. In the Link Model Editor, in "Supported link types," set "ptdup" to "yes" and all others to "no." This is to create a link model only in a point-to-point duplex version. Save this link model as "basic_link." In "Attributes" of this link, set "propdel model" to "dpt_propdel," set "txdel model" to "dpt_txdel," set "closure model," "coll model," "ecc model" and "error model" to "NONE." This is to make the link model's propagation delay and transmission delay follow point-to-point duplex models. However, "dpt_propdel" and "dpt_txdel" models belong to an external link delay file, and then it is required to declare this link delay file. To declare an external link delay file, from the Link Model Editor's "File" menu choose "Declare External Files..." and tick the "link_delay" checkbox. And then save this link model as "basic_link."

Q5.7 Why are there unresolved external symbol errors when running a simulation after adding a model into a project?

This is because this model may contain some files that are external to OPNET kernel. Therefore, one needs to declare them explicitly. To do that, in Project Editor, from "File" menu, choose "Declare External Files..." and tick the external files that may be used in your project.

Finally, a simulation project and a scenario can be created to test the custom node model. To do that, one can go through the following steps:

- From "File" menu, choose "New..." to create a new project. In the "Enter Name" dialog, set the "Project name" to "chapter5" and the "Scenario name" to "case1," and tick off the "Use Startup Wizard when creating new scenarios" checkbox.
- In Project Editor, press the "Open Object Palette" toolbar button. On the right side of the palette dialog, choose "Subnet" object and place this object on the Project Editor. This subnet is named "subnet_0", as in Figure 5.9.
- In "subnet_0," add two node objects to the Project Editor. The node model is "basic_source," which can be found in Object Palette Dialog.
- In Object Palette Dialog, find a link model called "basic_link." Connect the two "basic_source" node objects by this link object. The topology is shown in Figure 5.10.

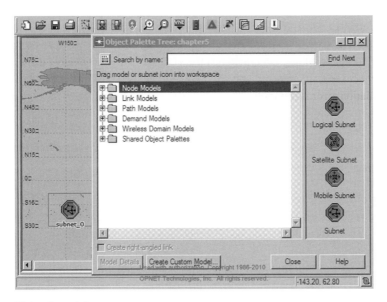

Figure 5.9 Network model

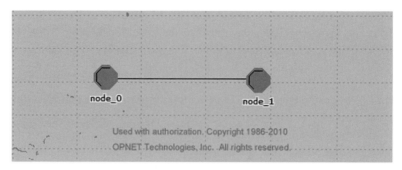

Figure 5.10　Network model

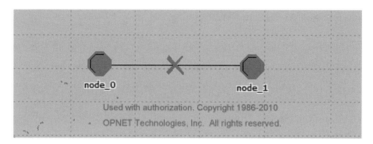

Figure 5.11　Network model

- From the "Topology" menu, choose "Verify Links..." to verify the connectivity between the link and nodes' transmitters and receivers. If there is a red cross over the link as in Figure 5.11, it means the link and nodes' transmitters and receivers are not connectable and some modifications are required to make them correctly connected. To correct connectivity errors, you can check (1) whether the link object's "data rate" attribute value matches the channel "data rate (bps)" attribute value for connected nodes' transmitters and receivers; (2) whether the link object's supported packet format matches the channel "packet formats" attribute of the connected nodes' transmitters and receivers; (3) whether the link object's "transmitter a," "receiver a," "transmitter b," and "receiver b" attributes have the correct values that should refer to the corresponding transmitters and receivers of connected nodes. Repeat this process until there are no connectivity errors.
- Choose some statistics of interest for the simulation kernel to collect. Right click the link object in Project Editor, select "Choose Individual DES Statistics." In the statistic results dialog, choose any statistic of interest for this link. In this case, "queuing delay (sec)" and "throughput (bits/sec)" are chosen in the "point-to-point" category.
- Now this simulation for this scenario can be run. In Project Editor, from the "DES" menu, choose "Configure/Run Discrete Event Simulation...." In the simulation configuration dialog, one may adjust some parameters. Then, press the "Run" button to start simulation.

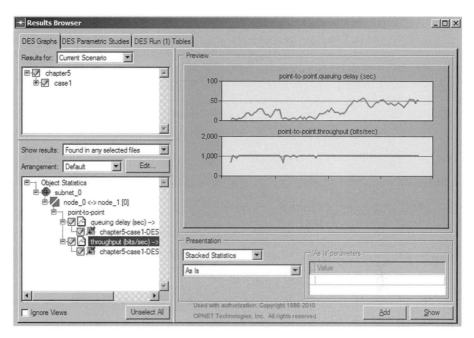

Figure 5.12   Results browser

- If there is no problem, the simulation will complete normally. Then, check the statistic results collected during simulation. In Project Editor, from the "DES" menu, choose "Results" – "View Results..." to show "Results Browser" (one can also right click any empty space in Project Editor, in the Context menu, and choose "View Results" to show "Results Browser"). In "Results Browser," tick the statistic results that are of interest. The statistic results will be shown in the "Results Browser" as in Figure 5.12. The results can be compared in "Results Browser." They can also be presented in different forms of distributions.
- To further process the statistic results via more advanced toolkits such as Excel and MATLAB, the statistic results can be exported to datasheet via OPNET Modeler's user interface. To do that, in "Results Browser," choose the statistics that are of interest, and press the "Show" button to show the results in a separate dialog. Right click the mouse on this separate dialog and choose "Export All Graph Data to Spreadsheet" to export the selected statistic results to a spreadsheet. This is shown in Figure 5.13.

Q5.8   Why does simulation output an "unsupported format" error?

This might be because the chosen link, transmitter and receiver data rates, or/and supported packet formats do not match, causing a connectivity error. One should check connectivity in Project Editor: from menu "Topology," choose "Verify Links" to see if there are connectivity errors.

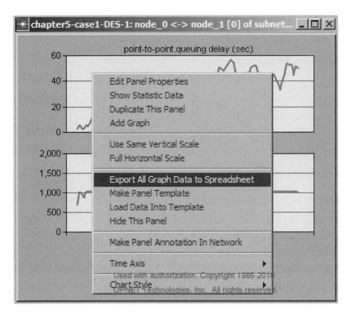

Figure 5.13   Statistic results

Stat Name	Mode	Count	Description	Group	Capture Mode	Draw Style	Low Bound	
Throughput (bits/sec)	Single	N/A			bucket/default total/sum_time		0.0	disabled

Figure 5.14   Local statistics

5.2.2   Case 2

In Case 1, the existing statistics are collected and viewed. In Case 2, custom statistics will be created in process model. Custom statistics in a "traffic_source" process model are created. Open the "traffic_source" process model in Process Editor. From the "Interfaces" menu, choose "Local Statistics." In the statistics dialog, add a new statistic as shown in Figure 5.14.

It is noted that the "Capture mode" of this statistic is "bucket/default total/sum_time." This is because this statistic is calculated as the accumulated statistic value divided by the elapsed time. Figure 5.15 shows how this capture mode is configured.

Q5.9   What are the differences between local statistics and global statistics?

Local statistics can only be accessed in the current process model, while global statistics can be accessed not only by the current process model but by other process models as well.

In Process Editor, open SV block to add a state variable "throughput", which is the handle of the custom statistic, as shown in Figure 5.16.

Figure 5.15 Capture mode

Type	Name	
Stathandle	throughput	

Figure 5.16 SV block

```
throughput = op_stat_reg("Throughput␣(bits/sec)", \
    OPC_STAT_INDEX_NONE , OPC_STAT_LOCAL);
```

Figure 5.17 "init" state

```
pk = op_pk_get(UPPER_IN_STRM_INDEX);
op_stat_write(throughput, op_pk_total_size_get(pk));
op_pk_send(pk, LOWER_OUT_STRM_INDEX);
```

Figure 5.18 "from_src" state

In "init" state, add code to register the custom statistic as in Figure 5.17. Note that the statistic name to register should be exactly the same as that set in the Local Statistics dialog, i.e., "Throughput (bits/sec)" in this case.

In "from_src" state, add code to record the throughput statistic, as in Figure 5.18. It is shown that the op_stat_write() function records the current received packet size in bits into the statistic. For the throughput statistic, its "Capture Mode" was previously set to "bucket/default total/sum_time." Therefore, the recorded packet sizes will be summed and divided by time elapsed to generate the throughput statistic.

Next, a new project scenario will be created and this custom statistic will be chosen before running the simulation. This new scenario is based on the scenario in Case 1. Open "chapter5-case1" scenario in Project Editor from "Scenarios" menu, choose "Duplicate

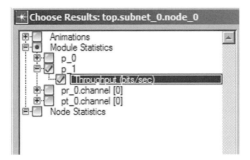

Figure 5.19   "Choose Results" dialog

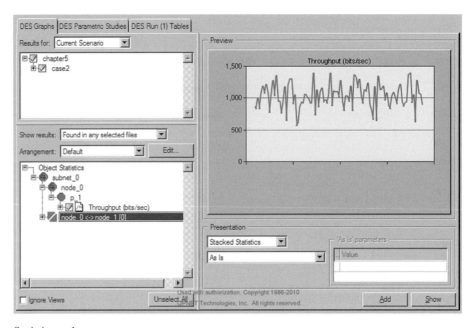

Figure 5.20   Statistic results

Scenario...." Name the scenario "case2." Right click "node_0," select "Choose Individual DES Statistics." In the "Choose Results" dialog, tick the "Throughput (bits/sec)" statistic of "p_1" module, as shown in Figure 5.19.

Now this simulation can be started from the "DES" menu. Alternatively, simply press "Ctrl+R" and "Alt+R" consecutively to start the simulation. After this simulation finishes, the custom statistic "Throughput (bits/sec)" can be viewed in "Results Browser", as shown in Figure 5.20.

5.2.3   Case 3

Both node model and modules have some predefined attributes. For node model, the predefined attributes can be "name," "model," "x position," "y position," etc. For modules within node model, the predefined attributes can be "name," "process model," "icon name," "begsim intrpt," etc. In Case 3, we will demonstrate how to add custom attributes both to modules and the containing node model, and how to access these attributes by code in process model.

5.2.3.1 Adding custom attribute to modules and the containing node model

Open "basic_source" node model in Node Editor. Right click "p_1" module, choose "Edit Attributes." In "(p_1) Attributes" dialog, press the "Extended Attrs." button to add attributes. In the "Extended Attributes" dialog, add a custom attribute as shown in Figure 5.21.

Now, "p_1" module has an extended custom attribute called "printed", as shown in Figure 5.22.

The initial value of the "printed" attribute is "promoted." If the value of the module's attribute is "promoted," this attribute will be exposed to the containing node model

Figure 5.21   Custom attributes

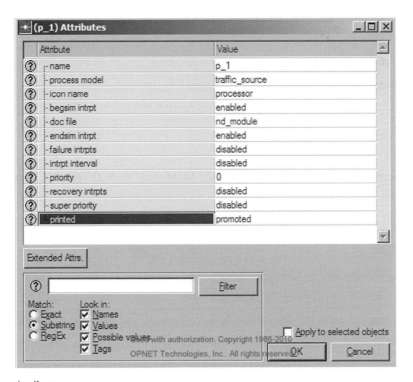

Figure 5.22   Attributes

as well, i.e., both "p_1" module and "basic_source" node model will have the custom attribute "printed." However, if you want to expose this attribute in "p_1" module only, you can set the value of the "printed" attribute to any integer number. To add a custom attribute to a node object, alternatively you can extend the node object's attribute directly by pressing the "Extended Attr." button in the node model's Attributes Dialog.

5.2.3.2   Accessing attributes by code in process model

Next, we will access the attributes by code in the process model. Double-click the "p_1" module to open the "traffic_source" process model in Process Editor. In SV block, add two new state variables, as shown in Figure 5.23.

In "init" state, add the code to get node object ID and store it in the "node_id" state variable, as shown in Figure 5.24.

In TV block, add two temporary variables as shown in Figure 5.25.

In "from_src" state, add the code shown in Figure 5.26 to implement such functionality: if the containing node model has "printed" attribute and if the value of this attribute is greater than 0, then print out the number of received packets in the simulation message console.

Save the "traffic_source" process model and compile it. In Project Editor, open chapter5 project and case2 scenario. Duplicate "chapter5-case2" project scenario and name it case3. Open the attribute dialog for "node_0" as in Figure 5.27; set attribute "p_1.printed" to an integer value greater than 0 if you want to print out the packet number, otherwise, set it to 0.

Type	Name	
Stathandle	throughput	
int	pk_num	
Objid	node_id	

Figure 5.23   SV block

```
throughput = op_stat_reg("Throughput␣(bits/sec)", \
     OPC_STAT_INDEX_NONE, OPC_STAT_LOCAL);

node_id = op_topo_parent(op_id_self());
```

Figure 5.24   "init" state

```
Packet *pk = OPC_NIL;
int printed = 0;
char msg[128];
```

Figure 5.25   TV block

```
pk = op_pk_get(UPPER_IN_STRM_INDEX);
op_stat_write(throughput, op_pk_total_size_get(pk));
if(op_ima_obj_attr_exists(node_id, "p_1.printed") \
    == OPC_TRUE)
{
  if(op_ima_obj_attr_get(node_id, "p_1.printed", \
      &printed) == OPC_COMPCODE_SUCCESS)
  {
    if(printed > 0)
    {
      sprintf(msg, "Total packet number: %d", \
          ++pk_num);
      op_sim_message( \
          "Custom attribute <printed> is ON", msg);
    }
  }
}
op_pk_send(pk, LOWER_OUT_STRM_INDEX);
```

Figure 5.26   "from_src" state

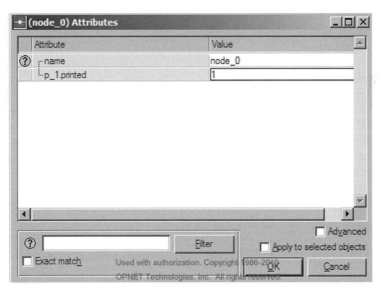

Figure 5.27   Attributes

Press "Ctrl+R" and "Alt+R" consecutively to start simulation. If "p_1.printed" attribute is set to a value greater than 0, the simulation message console will print out the packet number every time a packet is received in "node_0" node's "p_1" module, as shown in Figure 5.28.

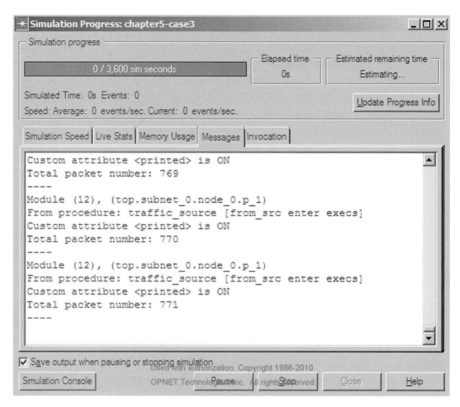

Figure 5.28   Simulation console

5.2.4   Case 4

In previous case studies, the packets had "NONE" packet format, i.e., unformatted packets. These packets are generated in the "p_0" module of the "basic_source" node model.
You can also modify the corresponding attributes of the "p_0" module to generate either
unformatted packets or formatted packets, and change packet size and interarrival times,
etc, as shown in Figure 5.29.

In Case 4, we will demonstrate how to create and access the formatted packet in the
process model, and how to create and use function in the process model. To create a
formatted packet, you need Packet Format Editor. In "File" menu, choose "New..." to
create a new "Packet Format." In the Packet Format Editor, press the "Create New Field"
toolbar button to place two fields on the Packet Format Editor, as shown in Figure 5.30.

Edit the attributes of these two fields. For the first field, set the "name" attribute to
"header" and set the "type" attribute to "integer." For the second field, set the "name"
attribute to "payload," set the "type" attribute to "packet" and set the size to "inherited."
Setting a field size to "inherited" will allow this field to have the actual data size. For
example, if this field is a packet field and the packet put into this field is 1024 bits, then
this field size is 1024 bits. If the field size is not "inherited," then the field always has
the size you set regardless of the actual data size in this field. You can add more fields if
you need the packet to carry more information. You may also play with other attributes

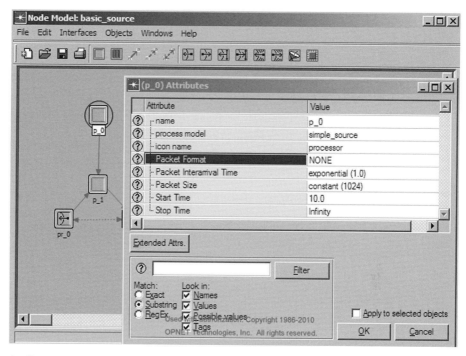

Figure 5.29   Attributes

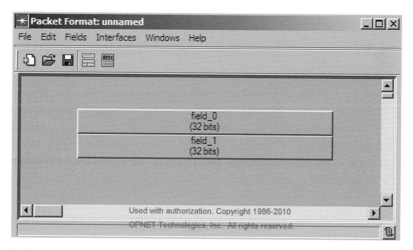

Figure 5.30   Packet Format Editor

of these fields, like the "color" attribute. Save this packet format as "wrapper_pk." The final created formatted packet is shown in Figure 5.31.

Open the "traffic_source" process model in Process Editor. In SV block, add a state variable that is a distribution handle used to generate a random number for the header. This is shown in Figure 5.32.

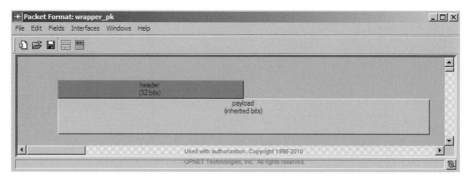

Figure 5.31   Packet Format Editor

Type	Name	
Stathandle	throughput	
int	pk_num	
Objid	node_id	
Distribution *	header_dist	

Figure 5.32   SV block

```
throughput = op_stat_reg("Throughput␣(bits/sec)", \
    OPC_STAT_INDEX_NONE, OPC_STAT_LOCAL);


node_id = op_topo_parent(op_id_self());


header_dist = op_dist_load("uniform_int", 0, 10);
```

Figure 5.33   "init" state

```
Packet *pk = OPC_NIL;
Packet *wrapper_pk = OPC_NIL;
int printed = 0;
int header = 0;
char msg[128];
```

Figure 5.34   TV block

In "init" state, add code to load a uniform distribution between integer 0 and 10, shown in Figure 5.33.

In TV block, add declarations of two temporary variables: "wrapper_pk" and "header," shown in Figure 5.34.

In HB block, add declaration of a function called "print_pk_size," which is used to print out packet size, shown in Figure 5.35. The function is set to static so that the

```
#define UPPER_IN_STRM_INDEX 1
#define LOWER_IN_STRM_INDEX 0
#define LOWER_OUT_STRM_INDEX 0
#define UPPER_STRM ( \
    (op_intrpt_type() == OPC_INTRPT_STRM) && \
    (op_intrpt_strm() == UPPER_IN_STRM_INDEX))
#define LOWER_STRM ( \
    (op_intrpt_type() == OPC_INTRPT_STRM) && \
    (op_intrpt_strm() == LOWER_IN_STRM_INDEX))

static void print_pk_size(Packet *);
```

Figure 5.35 HB block

```
static void print_pk_size(Packet *pk)
{
  char msg[128];

  FIN(print_pk_size(pk));

  sprintf(msg, "Payload packet size: %d", \
      op_pk_total_size_get(pk));
  op_sim_message("Wrapper packet header > 5", msg);

  FOUT;
}
```

Figure 5.36 FB block

function is valid only in the current source file and there is no naming conflict if other process model files define a function having the same name.

Q5.10 Why do we use "static" keywords?

This is because different process models are saved as different source files. A "static" keyword defined for a function is used to limit the function in the scope of the current source file of the simulation program. In other source files of the program, functions with the same name can be defined without a "redefinition" error.

In FB block, add code to implement the function, shown in Figure 5.36. The macros "FIN" and "FOUT" are used in OPNET functions to enable the OPNET debugging kernel to print out function information. For more details, check Chapter 11 on the topics of debugging in OPNET Modeler.

In "from_src" state, add code to create the formatted packet "wrapper_pk," set the "header" field of the formatted packet to a random integer number from 0 to 10, and

```
pk = op_pk_get(UPPER_IN_STRM_INDEX);
op_stat_write(throughput, op_pk_total_size_get(pk));
if(op_ima_obj_attr_exists(node_id, "p_1.printed") \
     == OPC_TRUE)
{
  if(op_ima_obj_attr_get(node_id, "p_1.printed", \
        &printed) == OPC_COMPCODE_SUCCESS)
  {
    if(printed > 0)
    {
      sprintf(msg, "Total_packet_number:_%d", ++pk_num);
      op_sim_message( \
           "Custom_attribute_<printed>_is_ON", msg);
    }
  }
}
wrapper_pk = op_pk_create_fmt("wrapper_pk");
op_pk_nfd_set(wrapper_pk, "header", \
     (int)op_dist_outcome(header_dist));
op_pk_nfd_set(wrapper_pk, "payload", pk);
op_pk_send(wrapper_pk, LOWER_OUT_STRM_INDEX);
```

Figure 5.37   "from_src" state

```
wrapper_pk = op_pk_get(LOWER_IN_STRM_INDEX);
op_pk_nfd_get(wrapper_pk, "header", &header);
op_pk_nfd_get(wrapper_pk, "payload", &pk);
if(header > 5)
  print_pk_size(pk);
op_pk_destroy(pk);
op_pk_destroy(wrapper_pk);
```

Figure 5.38   "from_recv" state

set the "payload" field to the unformatted packet received from the upper layer. This is shown in Figure 5.37.

In "from_recv" state, add code to get the fields of the received formatted packet. If the value of the "header" field is greater than 5, then print the size of the payload packet. Finally, destroy both payload packet and wrapper packet. This is shown in Figure 5.38.

Save and compile the "traffic_source" model. In Project Editor, open chapter5 project and case3 scenario. Duplicate "chapter5-case3" project scenario and name it case4. Press "Ctrl+R" and "Alt+R" consecutively to start simulation. During simulation, the payload packet size will be printed out if the wrapper packet's header value is greater than 5, as shown in Figure 5.39.

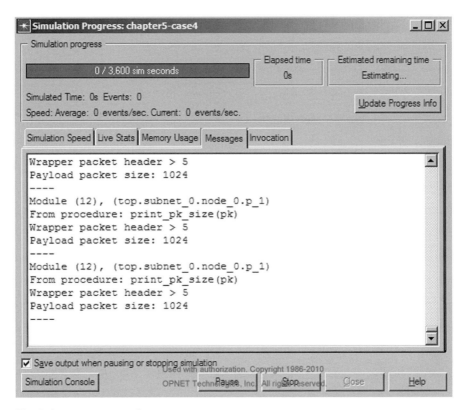

Figure 5.39   Simulation message console

5.2.5   Case 5

In this case, we will create a node that is capable of switching packets to different ports.
The node is able to switch packets at a certain rate. Incoming packets will first be pushed
into a buffer, and packets in the buffer will be removed and sent out at a certain switching
rate. If the incoming packet rate is greater than the switching rate, then packets will be
queued in the buffer until the incoming packet rate is reduced. First, we will create a
new node model called "pk_switch." This node has five pairs of transmitter/receiver
connected to a queue module, as shown in Figure 5.40.

Q5.11   What are the differences between processor module and queue module?

Both processor module and queue module can be used as processor. However, for
the queue module, it is also able to access queuing facilities via OPNET Queue and
Subqueue API packages.

Now, you need to create a process model for the "q_0" module so that it is capable of
switching packets. Create a new process model called "pk_switch"; the state transition
diagram is shown in Figure 5.41.

In SV block, declare a state variable for storing a timeout event handle, shown in
Figure 5.42.

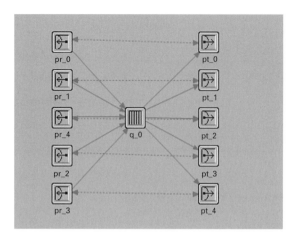

Figure 5.40   Custom node model

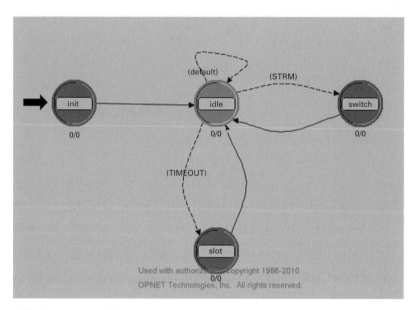

Figure 5.41   STD of process model

In TV block, add two temporary variables, shown in Figure 5.43.

In HB block, add macro definitions for "STRM" and "TIMEOUT" state transition conditions. "STRM" is defined as packet stream interrupt. "TIMEOUT" is defined as self-interrupt, which is scheduled by user code. "SLOT_DURATION" is the interval between scheduled self-interrupts. "PORT_NUM" is the number of output ports that the "q_0" module has. This is shown in Figure 5.44.

In "init" state, add code to schedule the first self-interrupt that will trigger the slot state for removing the packet in the buffer. This is shown in Figure 5.45.

Type	Name	
Evhandle	timeout_handle	

Figure 5.42   SV block

```
Packet *pk = OPC_NIL;
int header = 0;
```

Figure 5.43   TV block

```
#define STRM (op_intrpt_type() == OPC_INTRPT_STRM)
#define TIMEOUT_INTRPT_CODE 0
#define TIMEOUT ( \
    (op_intrpt_type() == OPC_INTRPT_SELF) && \
    (op_intrpt_code() == TIMEOUT_INTRPT_CODE))
#define SLOT_DURATION 0.2
#define PORT_NUM 5
```

Figure 5.44   HB block

```
timeout_handle = op_intrpt_schedule_self( \
    op_sim_time() + SLOT_DURATION, TIMEOUT_INTRPT_CODE);
```

Figure 5.45   "init" state

```
pk = op_pk_get(op_intrpt_strm());
op_subq_pk_insert(0, pk, OPC_QPOS_TAIL);
```

Figure 5.46   "switch" state

In "STRM" state, add code to push the incoming packet to the end of the first subqueue, shown in Figure 5.46. To access a buffer via OPNET queue or subqueue APIs, you must use a queue module instead of a processor module. In this case, we use a queue module "q_0." Make sure the "subqueue" attribute of "q_0" module has at least one subqueue row as shown in Figure 5.47, otherwise the API used to access the first subqueue will fail. If you want to access more subqueues, you need to add more rows in the "subqueue" attribute table of "q_0" module.

In "slot" state, add code to remove the packet from the first subqueue and schedule another self-interrupt at the next time slot, shown in Figure 5.48.

The process model "pk_switch" models a leaky-bucket scenario. Incoming packets will be pushed into a buffer. Self-interrupt is scheduled at an interval for the slot to remove packet in the buffer. The removal rate is fixed and is limited by "SLOT_DURATION." Save and compile "pk_switch" process model.

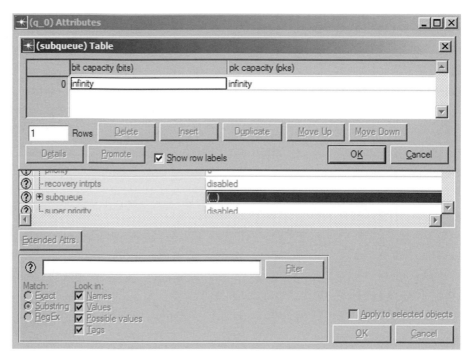

Figure 5.47   Subqueue table

```
if(op_ev_valid(timeout_handle))
{
  if(op_subq_empty(0) == OPC_FALSE)
  {
    pk = op_subq_pk_remove(0, OPC_QPOS_HEAD);
    op_pk_nfd_access(pk, "header", &header);
    op_pk_send(pk, header%PORT_NUM);
  }
  timeout_handle = op_intrpt_schedule_self( \
      op_sim_time() + SLOT_DURATION, \
      TIMEOUT_INTRPT_CODE);
}
```

Figure 5.48   "slot" state

Next, we will create a project scenario for simulation. In Project Editor, open chapter5 project and case4 scenario. Duplicate "chapter5-case4" project scenario and name it case5. Create a network that is the same as in Figure 5.49. "switch" has "pk_switch" node model type. "node_0" to "node_4" have "basic_source" node model type. Choose "subqueue" statistics for "switch" node object. Press "Ctrl+R" and "Alt+R" consecutively to start simulation. After simulation finishes, view the "subqueue" statistic results for "switch" node object. Figure 5.49 shows the results.

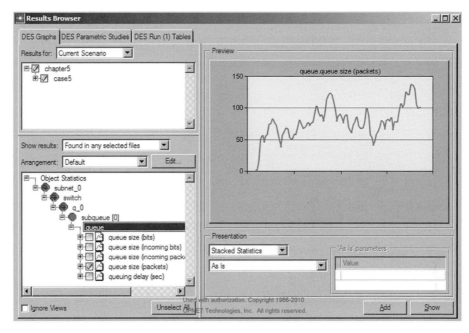

Figure 5.49   Statistic results

Next, we will increase the packet rate for the traffic source node to see how switch node performs. Double-click any of the "node_0" to "node_4" to open "basic_source" node model in Node Editor. Set the "Packet Interarrival Time" attribute of the "p_0" module to 0.5. Save the "basic_source" node model. Run the simulation again. After simulation finishes, view the "subqueue" statistic results for "switch" node object. Figure 5.50 shows the results.

It is seen that after increasing the packet rate for traffic source nodes, the switch is not fast enough to remove the packets in the buffer, so the buffer of the switch keeps growing. You can also reduce the value of "SLOT_DURATION" in "pk_switch" process model to increase the switch node's switching rate.

5.2.6   Case 6

In Case 5, we showed how to switch packets without checking any routing table. In Case 6, we will demonstrate how to build a routing table and how to apply a shortest path routing algorithm to our model.

The graph is the base for a building routing table. First, we will show how to construct a graph for a particular network topology. A network topology can be modeled as a graph. Nodes can be modeled as vertices and links can be modeled as edges in the graph. If the links are simplex, the graph can be regarded as a directed graph. In OPNET API, there is a Graph package which can be used to model the network topology.

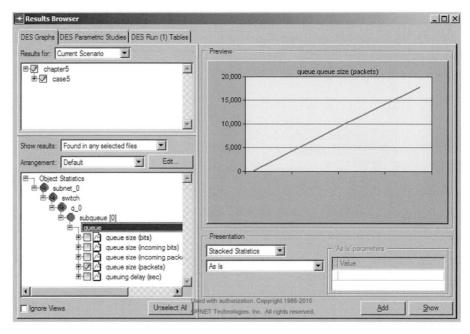

Figure 5.50   Statistic results

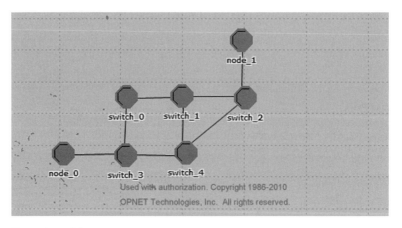

Figure 5.51   Network model

In Project Editor, open chapter5 project and case5 scenario. Duplicate "chapter5-case5" project scenario and name it case6. Create a network that is the same as in Figure 5.51; "switch_0" to "switch_4" have "pk_switch" node model type; "node_0" and "node_1" have "basic_source" node model type.

Among these nodes, "switch_0" to "switch_4" are capable of routing and "node_0" and "node_1" are end-to-end traffic source nodes. The simulation task is to allow traffic sent from "node_0" to be routed to "node_1" and traffic sent from "node_1" to be routed to "node_0." The actual routing graph is shown in Figure 5.52.

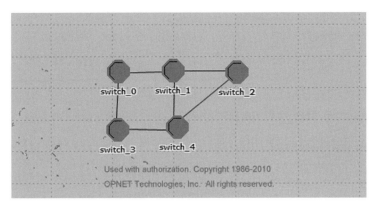

Figure 5.52   Actual routing graph

Type	Name	
Evhandle	timeout_handle	
PrgT_Graph *	graph	
PrgT_Graph_State_ID	graph_state_id	
Objid	switch_nodes[7]	
PrgT_Graph_Vertex *	vertices[7]	
PrgT_Graph_Edge *	edges[8]	
Objid	node_id	
Objid	mod_id	
PrgT_List *	path_nodes_id_list	

Figure 5.53   SV block

First, this routing graph will be built by using Graph APIs. Open the "pk_switch" process model in Process Editor. In SV block, add declarations of several state variables as shown in Figure 5.53.

In TV block, add declarations of several temporary variables, shown in Figure 5.54.

In HB block, include header files for graph, routing, and vector API packages, and add macro definitions for graph namespace and graph state, as shown in Figure 5.55.

Next, you will build a graph for the network topology shown in Figure 5.51. The graph is built in "init" state, shown in Figures 5.56 and 5.57.

In Figures 5.56 and 5.57, you first register a state handler with a defined graph namespace name and state name, and then create an empty graph with the graph namespace name. Namespace names allow creation of different dimensions of graphs. The returned "graph_state_id" and "graph" objects can be used to access other graphs and routing APIs. Then, you can retrieve all object IDs of the nodes in the network. These object IDs will be associated with corresponding vertices in the graph. Next, you can add vertices and edges into the graph. Especially for vertices, you can associate them with the corresponding node's object ID. Now, the "graph" object contains all topology information for the network shown in Figure 5.51.

```
Packet *pk = OPC_NIL;
int header = 0;
int i = 0;
int j = 0;
int count = 0;
int count2 = 0;
Boolean found = FALSE;
Objid strm_id;
Objid tx_id;
Objid link_id;
Objid next_hop_node_id;
Objid tmp_node_id;
int out_port = 0;
PrgT_Vector *shortest_paths;
PrgT_Vector *shortest_path;
int      path_nodes_size;
PrgT_Vector *path_nodes;
Objid *temp_id_ptr = OPC_NIL;
```

Figure 5.54   TV block

```
#include <prg.h>
#include <prg_vector.h>
#include <prg_graph.h>
#include <prg_djk.h>
#define STRM (op_intrpt_type() == OPC_INTRPT_STRM)

#define TIMEOUT_INTRPT_CODE 0
#define TIMEOUT ( \
    (op_intrpt_type() == OPC_INTRPT_SELF) && \
    (op_intrpt_code() == TIMEOUT_INTRPT_CODE))

#define SLOT_DURATION 0.2
#define PORT_NUM 5

#define GRAPH_NAMESPACE_NAME "graph_namespace"
#define GRAPH_STATE_NAME "graph_state_name"
```

Figure 5.55   HB block

In this simulation scenario, we want to route traffic from "node_0" to "node_1" and in the reverse direction as well. Since there are several different ways to route traffic between these two nodes, you need to find the shortest path to do that. In OPNET, the DJK package provides APIs that are capable of finding a set of shortest paths from a source vertex to a destination vertex in a graph. "Shortest path" is the least-cost

```
graph_state_id = prg_graph_state_handler_register( \
    GRAPH_NAMESPACE_NAME, GRAPH_STATE_NAME, \
    PRGC_NIL, PRGC_NIL);
graph = prg_graph_create(GRAPH_NAMESPACE_NAME);
// Get nodes' id
switch_nodes[0] = op_id_from_hierarchical_name( \
    "top.subnet_0.switch_0");
switch_nodes[1] = op_id_from_hierarchical_name( \
    "top.subnet_0.switch_1");
switch_nodes[2] = op_id_from_hierarchical_name( \
    "top.subnet_0.switch_2");
switch_nodes[3] = op_id_from_hierarchical_name( \
    "top.subnet_0.switch_3");
switch_nodes[4] = op_id_from_hierarchical_name( \
    "top.subnet_0.switch_4");
switch_nodes[5] = op_id_from_hierarchical_name( \
    "top.subnet_0.node_0");
switch_nodes[6] = op_id_from_hierarchical_name( \
    "top.subnet_0.node_1");
// Add vertices to graph
// Associate vertices with nodes' id
vertices[0] = prg_graph_vertex_insert(graph);
prg_vertex_client_state_set(vertices[0], \
    graph_state_id, &switch_nodes[0]);
vertices[1] = prg_graph_vertex_insert(graph);
prg_vertex_client_state_set(vertices[1], \
    graph_state_id, &switch_nodes[1]);
vertices[2] = prg_graph_vertex_insert(graph);
prg_vertex_client_state_set(vertices[2], \
    graph_state_id, &switch_nodes[2]);
vertices[3] = prg_graph_vertex_insert(graph);
prg_vertex_client_state_set(vertices[3], \
    graph_state_id, &switch_nodes[3]);
vertices[4] = prg_graph_vertex_insert(graph);
prg_vertex_client_state_set(vertices[4], \
    graph_state_id, &switch_nodes[4]);
vertices[5] = prg_graph_vertex_insert(graph);
prg_vertex_client_state_set(vertices[5], \
    graph_state_id, &switch_nodes[5]);
vertices[6] = prg_graph_vertex_insert(graph);
prg_vertex_client_state_set(vertices[6], \
    graph_state_id, &switch_nodes[6]);
```

Figure 5.56   "init" state

```
// Add edges to graph
edges[0] = prg_graph_edge_insert(graph, vertices[0], \
    vertices[1], PrgC_Graph_Edge_Duplex);
edges[1] = prg_graph_edge_insert(graph, vertices[0], \
    vertices[3], PrgC_Graph_Edge_Duplex);
edges[2] = prg_graph_edge_insert(graph, vertices[1], \
    vertices[2], PrgC_Graph_Edge_Duplex);
edges[3] = prg_graph_edge_insert(graph, vertices[1], \
    vertices[4], PrgC_Graph_Edge_Duplex);
edges[4] = prg_graph_edge_insert(graph, vertices[2], \
    vertices[4], PrgC_Graph_Edge_Duplex);
edges[5] = prg_graph_edge_insert(graph, vertices[3], \
    vertices[4], PrgC_Graph_Edge_Duplex);
edges[6] = prg_graph_edge_insert(graph, vertices[5], \
    vertices[3], PrgC_Graph_Edge_Duplex);
edges[7] = prg_graph_edge_insert(graph, vertices[6], \
    vertices[2], PrgC_Graph_Edge_Duplex);
```

Figure 5.57   "init" state

path, where cost is the sum of the edge weights of a path. You can set the weights of edges in a graph to model the actual cost. Weight in a graph is a relative value. For the graph defined for the network in Figure 5.51, you can set the following weights for the edges:

- Edge for "switch_0" and "switch_1": 1
- Edge for "switch_0" and "switch_3": 1
- Edge for "switch_1" and "switch_2": 1
- Edge for "switch_1" and "switch_4": 0.2
- Edge for "switch_2" and "switch_4": 1.5
- Edge for "switch_3" and "switch_4": 1
- Edge for "node_0" and "switch_3": 1
- Edge for "node_1" and "switch_2": 1

From manual calculation, the shortest path between "node_0" and "node_1" should be "switch_3" – "switch_4" – "switch_1" – "switch_2." The weight sum for this path is 4.2. In "init" state, add code to set weights for edges, compute the routing table, and get the shortest path between "node_0" and "node_1," shown in Figure 5.58.

In Figure 5.58, "path_nodes_id_list" is a list containing IDs of objects of the shortest path.

In "slot" state, add code to check the next hop node, based on the computed shortest path and switch packet to the corresponding output port connected to the next hop node, as shown in Figures 5.59 and 5.60.

Save and compile the process model.

```
prg_djk_graph_init(graph);

// Set weight for all edges
prg_djk_edge_weight_set(edges[0],  1);
prg_djk_edge_weight_set(edges[1],  1);
prg_djk_edge_weight_set(edges[2],  1);
prg_djk_edge_weight_set(edges[3],  0.2);
prg_djk_edge_weight_set(edges[4],  1.5);
prg_djk_edge_weight_set(edges[5],  1);
prg_djk_edge_weight_set(edges[6],  1);
prg_djk_edge_weight_set(edges[7],  1);
prg_djk_all_sources_compute(graph);

// Get all shortest paths from vertices[5] to vertices[6]
shortest_paths = prg_djk_path_get(vertices[5], \
    vertices[6]);
// Get the first shortest path of all shortest paths
shortest_path = \
    (PrgT_Vector *)prg_vector_access(shortest_paths, 0);
// Get all vertices for this shortest path
path_nodes = \
    prg_graph_path_vector_to_node_vector_create( \
    shortest_path, vertices[5], vertices[6]);
// Get all nodes' id associated with these vertices
path_nodes_size = prg_vector_size(path_nodes);
path_nodes_id_list = prg_list_create();
for(i = 0; i < path_nodes_size; ++i)
{
  temp_id_ptr = (int *)prg_vertex_client_state_get( \
      (PrgT_Graph_Vertex *)prg_vector_access( \
      path_nodes, i), graph_state_id);
  prg_list_insert(path_nodes_id_list, temp_id_ptr, \
      PRGC_LISTPOS_TAIL);
}
// Get current module id and node id
mod_id = op_id_self();
node_id = op_topo_parent(mod_id);
```

Figure 5.58 "init" state

Q5.12 The "wrapper_pk" packet's "header" field is int type. Why can Objid type variable be assigned to it?

This is because Objid type is internally defined as int type in the "p_objid_type.h" header file, which is located in the [OPNET Home]\sys\include directory.

```
if(op_ev_valid(timeout_handle) == OPC_FALSE)
  return;

if(op_subq_empty(0) == OPC_TRUE)
  return;

pk = op_subq_pk_remove(0, OPC_QPOS_HEAD);
// "header" contains the destination node id
op_pk_nfd_access(pk, "header", &header);
count = prg_list_size(path_nodes_id_list);
// Check if destination node id is at the top or
// at the tail of shortest path nodes list
if(*(int *)prg_list_access(path_nodes_id_list, \
    PRGC_LISTPOS_TAIL) == header)
{
  // If destination node id is at the tail,
  // the next hop node id is 1 after current node id
  for(i = 0; i < count; ++i)
  {
    if(*(int *)prg_list_access( \
        path_nodes_id_list, i) == node_id)
    {
      next_hop_node_id = *(int *)prg_list_access( \
          path_nodes_id_list, i + 1);
      break;
    }
  }
}
else // Destination node id is at the top of id list
{
  // If destination node id is at the top,
  // the next hop node id is 1 before current node id
  for(i = count - 1; i >= 0; --i)
  {
    if(*(int *)prg_list_access( \
        path_nodes_id_list, i) == node_id)
    {
      next_hop_node_id = *(int *)prg_list_access( \
          path_nodes_id_list, i - 1);
      break;
    }
  }
}
```

Figure 5.59   "slot" state

```
count = op_topo_assoc_count(mod_id, \
    OPC_TOPO_ASSOC_OUT, OPC_OBJTYPE_STRM);
found = FALSE;
// Check all surrounding modules to see which one is
// connected to next hop node. If a module is found
// to be able to connect to next hop node, retrieve
// the output stream port leading to that module
for(i = 0; i < count; ++i)
{
  strm_id = op_topo_assoc(mod_id, OPC_TOPO_ASSOC_OUT, \
      OPC_OBJTYPE_STRM, i);
  tx_id = op_topo_assoc(strm_id, OPC_TOPO_ASSOC_OUT, \
      OPC_OBJTYPE_PTTX, 0);
  if(op_topo_assoc_count(tx_id, OPC_TOPO_ASSOC_OUT, \
      OPC_OBJTYPE_LKDUP) == 0)
    continue;
  link_id = op_topo_assoc(tx_id, OPC_TOPO_ASSOC_OUT, \
      OPC_OBJTYPE_LKDUP, 0);
  count2 = op_topo_assoc_count(link_id, \
      OPC_TOPO_ASSOC_OUT, OPC_OBJTYPE_NODE_FIX);
  for(j = 0; j < count2; ++j)
  {
    tmp_node_id = op_topo_assoc(link_id, \
        OPC_TOPO_ASSOC_OUT, OPC_OBJTYPE_NODE_FIX, j);
      if(tmp_node_id == next_hop_node_id)
    {
      op_ima_obj_attr_get(strm_id, \
          "src stream", &out_port);
      found = TRUE;
      break;
    }
  }
  if(found == TRUE)
    break;
}
// Send packet to the found output port
op_pk_send(pk, out_port);
timeout_handle = op_intrpt_schedule_self( \
    op_sim_time() + SLOT_DURATION, TIMEOUT_INTRPT_CODE);
```

Figure 5.60 "slot" state

In Figure 5.59, the "header" field of the wrapper packet contains the object ID of the destined node. You should also modify the "traffic_source" process model to allow "node_0" and "node_1" nodes to send packets with "header" field containing the object ID of the destined node. Open the "traffic_source" process model in Process Editor. In

```
pk = op_pk_get(UPPER_IN_STRM_INDEX);
op_stat_write(throughput, op_pk_total_size_get(pk));
if(op_ima_obj_attr_exists(node_id, "p_1.printed") \
    == OPC_TRUE)
{
  if(op_ima_obj_attr_get(node_id, "p_1.printed", \
      &printed) == OPC_COMPCODE_SUCCESS)
  {
    if(printed > 0)
    {
      sprintf(msg, "Total packet number: %d", \
          ++pk_num);
      op_sim_message( \
          "Custom attribute <printed> is ON", msg);
    }
  }
}
// Create wrapper packet and set the "header" field
// of wrapper packet to the destination node id
// Note: the destination node id is obtained from
// destination node name which is specified in source
// node's "dest node name" attribute
wrapper_pk = op_pk_create_fmt("wrapper_pk");
if(op_ima_obj_attr_exists(node_id, "dest node name") \
    == OPC_TRUE)
{
  if(op_ima_obj_attr_get(node_id, "dest node name", \
      &dest_node_name) == OPC_COMPCODE_SUCCESS)
  {
    dest_node_id = op_id_from_hierarchical_name( \
        dest_node_name);
    op_pk_nfd_set(wrapper_pk, "header", dest_node_id);
  }
}
op_pk_nfd_set(wrapper_pk, "payload", pk);
op_pk_send(wrapper_pk, LOWER_OUT_STRM_INDEX);
```

Figure 5.61   "from_src" state

"from_src" state, add code to set the object ID of the destination node to the "header" field of the wrapper packet, shown in Figure 5.61. The object ID of the destination node is obtained from the name of the destination node, which is set in the custom attribute "dest node name" of "node_0" and "node_1" nodes.

Save and compile the process model.

For both "node_0" and "node_1," add a custom attribute "dest node name." For "node_0," set the value of "dest node name" attribute to "top.subnet_0.node_1." For

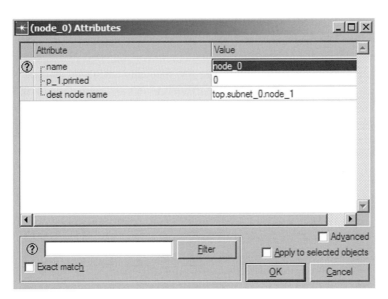

Figure 5.62   Attribute dialogue

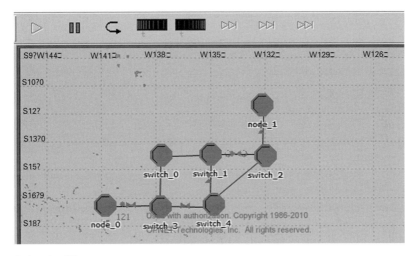

Figure 5.63   Animation Viewer

"node_1," set the value of "dest node name" attribute to "top.subnet_0.node_0." This is
shown in Figure 5.62.

We are now nearly ready to run the simulation scenario to see if these models work
correctly. The OPNET animation facility can be used to see how packet streams are
transferred in the subnet and to validate traffic routing. To record animation for the
simulation scenario, in Project Editor, from the "DES" menu choose "Record Packet
Flow 2D Animation For Subnet." Now, you can run the simulation for the case6 scenario.
After simulation finishes, from the "DES" menu, choose "Play 2D Animation" to show
Animation Viewer, which is shown in Figure 5.63.

From Figure 5.63, it is seen that the packet flow route is the same as the shortest path you manually calculated.

Next, you will model an error situation where the edge between "switch_1" and "switch_4" is unavailable. In this case, the manually calculated shortest path becomes "switch_3" – "switch_4" – "switch_2." To model this error situation, in "init" state of the "pk_switch" process model, use the "prg_djk_edge_disable()" function to disable the edge between "switch_1" and "switch_4," which is "edges[3]," as shown in Figure 5.64.

```
prg_djk_graph_init(graph);
prg_djk_edge_weight_set(edges[0], 1);
prg_djk_edge_weight_set(edges[1], 1);
prg_djk_edge_weight_set(edges[2], 1);
prg_djk_edge_weight_set(edges[3], 0.2);
prg_djk_edge_weight_set(edges[4], 1.5);
prg_djk_edge_weight_set(edges[5], 1);
prg_djk_edge_weight_set(edges[6], 1);
prg_djk_edge_weight_set(edges[7], 1);
// Disable edges[3]
prg_djk_edge_disable(edges[3]);
prg_djk_all_sources_compute(graph);

shortest_paths = prg_djk_path_get(vertices[5], \
    vertices[6]);
shortest_path = (PrgT_Vector *)prg_vector_access( \
    shortest_paths, 0);
path_nodes = \
    prg_graph_path_vector_to_node_vector_create( \
    shortest_path, vertices[5], vertices[6]);
path_nodes_size = prg_vector_size(path_nodes);
path_nodes_id_list = prg_list_create();
for(i = 0; i < path_nodes_size; ++i)
{
  temp_id_ptr = (int *)prg_vertex_client_state_get( \
      (PrgT_Graph_Vertex *)prg_vector_access( \
      path_nodes, i), graph_state_id);
  prg_list_insert(path_nodes_id_list, temp_id_ptr, \
      PRGC_LISTPOS_TAIL);
}

mod_id = op_id_self();
node_id = op_topo_parent(mod_id);
```

Figure 5.64   "init" state

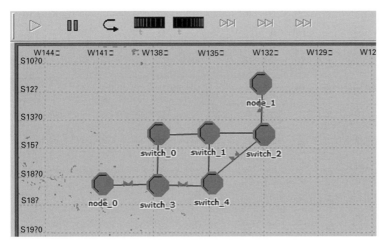

Figure 5.65 Animation Viewer

Save and compile the process model "pk_switch." Run the simulation scenario for case6 again. Figure 5.65 shows the animation results. It is shown that the alternative shortest path under such an error situation is the same as the manually calculated shortest path.

Animation of these scenarios can also be recorded into video for demonstration by using screen recording tools such as SMRecorder and Snagit.

5.2.7 Case 7

In Case 7, we will demonstrate how to carry extra information with packets. In case 4, you used the "header" field to carry information with the packet. It models the real packet's functions. However, sometimes you want a packet to carry some extra information that does not exist in the real packet and is only used to facilitate simulation control. For this purpose, you can use Interface Control Information (ICI) Package to manipulate the ICI object and associate it with a packet. ICI is a data type in OPNET. It can be used to carry integer, double, and structure types of data. ICI can also be associated with a packet so that the packet can carry extra control information.

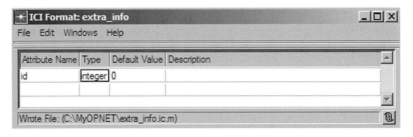

Figure 5.66 ICI format

Type	Name	
Stathandle	throughput	
int	pk_num	
Objid	node_id	
Distribution *	header_dist	
int	ici_id	

Figure 5.67   SV block

```
Packet *pk = OPC_NIL;
Packet *wrapper_pk = OPC_NIL;
int printed = 0;
int header = 0;
char msg[128];
char dest_node_name[128];
Objid dest_node_id;
Ici *ici = OPC_NIL;
int temp;
```

Figure 5.68   TV block

To use ICI, you need to create an ICI format first. From "File" menu, choose "New..." – choose "ICI Format." In the ICI Format Editor, add a new attribute and save this ICI format, as shown in Figure 5.66.

Now, you need to add code to see how to associate the ICI object with packets and how to access ICI objects. Open the "traffic_source" process model in Process Editor. In SV block, add a state variable as shown in Figure 5.67.

In TV block, add a temporary variable as shown in Figure 5.68.

In "from_src" state, add code to create an ICI object, set the attribute of the ICI object and associate it with a wrapper packet, shown in Figure 5.69.

In "from_recv" state, add code to get the ICI object associated with the wrapper packet, get the attribute value, and print this value and destroy the ICI object, shown in Figure 5.70.

Save and compile "traffic_source" process model. Now, you can run the simulation to see the output, as shown in Figure 5.71. However, you can use the ICI to carry any control information.

5.3   Model optimization and validation

Optimizing the model can make simulation run faster. Optimizing process model code is similar to optimizing general C/C++ code. In particular, if a value can be obtained at the beginning of simulation and will be used frequently during simulation invocations,

```
pk = op_pk_get(UPPER_IN_STRM_INDEX);
op_stat_write(throughput, op_pk_total_size_get(pk));
if(op_ima_obj_attr_exists(node_id, "p_1.printed") \
    == OPC_TRUE)
{
  if(op_ima_obj_attr_get(node_id, "p_1.printed", \
      &printed) == OPC_COMPCODE_SUCCESS)
  {
    if(printed > 0)
    {
      sprintf(msg, "Total␣packet␣number:␣%d", \
          ++pk_num);
      op_sim_message( \
          "Custom␣attribute␣<printed>␣is␣ON", msg);
    }
  }
}
wrapper_pk = op_pk_create_fmt("wrapper_pk");
if(op_ima_obj_attr_exists(node_id, "dest␣node␣name") \
    == OPC_TRUE)
{
  if(op_ima_obj_attr_get(node_id, "dest␣node␣name", \
      &dest_node_name) == OPC_COMPCODE_SUCCESS)
  {
    dest_node_id = op_id_from_hierarchical_name( \
        dest_node_name);
    op_pk_nfd_set(wrapper_pk, "header", dest_node_id);
  }
}
// Create ICI object, assign a value to its "id"
// attribute, and associate it with wrapper packet
op_pk_nfd_set(wrapper_pk, "payload", pk);
ici = op_ici_create("extra_info");
op_ici_attr_set_int32(ici, "id", ++ici_id);
op_pk_ici_set(wrapper_pk, ici);
op_pk_send(wrapper_pk, LOWER_OUT_STRM_INDEX);
```

Figure 5.69   "from_src" state

then you should get this value at the initialization stage and assign it to a state variable, and reference this state variable during simulation. Optimization of such frequently invoked states as stream interrupt states can make significant performance improvements. Further, if you are not going to debug your model during simulation, you can set the "Simulation Kernel Type" preference to "optimized." This preference can be found from the "Edit" menu – Preferences, as shown in Figure 5.72. Alternatively, you can

```
wrapper_pk = op_pk_get(LOWER_IN_STRM_INDEX);
// Get the ICI object associated with wrapper
// packet, print out the value of "id"
// attribute for this ICI object and destroy
// this object afterwards
ici = op_pk_ici_get(wrapper_pk);
op_ici_attr_get_int32(ici, "id", &temp);
sprintf(msg, "ICI id: %d", temp);
op_sim_message("ICI test", msg);
op_ici_destroy(ici);
op_pk_nfd_get(wrapper_pk, "header", &header);
op_pk_nfd_get(wrapper_pk, "payload", &pk);
if(header > 5)
   print_pk_size(pk);
op_pk_destroy(pk);
op_pk_destroy(wrapper_pk);
```

Figure 5.70   "from_recv" state

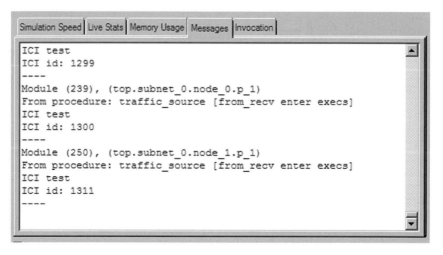

Figure 5.71   Simulation message console

set "Simulation Kernel" to "Optimized" in simulation configuration dialog as shown in Figure 5.73.

There is no perfect method for model validation. However, a good practice to validate a model is to validate progressively throughout the model creation process, i.e., validate the model whenever a new functionality is added to it. The validation process can be repetitive. Here are some methods that can be applied during the model validation process: check and compare simulation statistic results to see if they match common sense, experience, or are close to mathematical results if a math model is available; debug

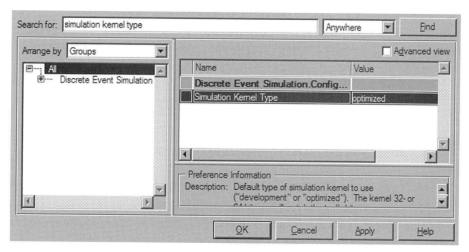

Figure 5.72   Preferences dialogue

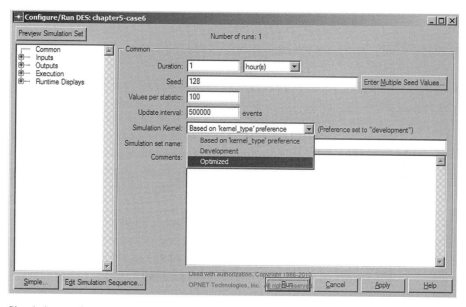

Figure 5.73   Simulation configuration dialog

the model repetitively by applying debugging techniques discussed in Chapter 11; use animation to validate models since animation shows how packets flow between nodes in the subnet and modules within a node; compare simulation results with experimental results.

6 High-level wrapper APIs

This chapter demonstrates how to write your own high-level wrapper APIs which encapsulate many OPNET programming details, in order to facilitate and accelerate design and programming of OPNET models. We also provide several wrapper API packages to help users to quickly build their models. To be able to follow this chapter, you are assumed to know how to write custom models as shown in Chapter 5, and the basics of generic programming.

6.1   Why and how to use wrapper APIs

When creating a simulation model, one often needs to repetitively write code with some similar functionalities. To speed up modeling and coding, one can write some wrapper functions that implement those commonly used functionalities. These wrapper functions will save modeling time and reduce the chances of making errors. Wrapper APIs here refer to the functions that encapsulate many programming details and implement particular functionalities at higher level. Figure 6.1 shows an example of a wrapper API that is defined in a header file called "geo_topo.h", which can be downloaded from the publisher's website.

Figure 6.1 shows that a wrapper API is simply a function that wraps some functionalities and provides a higher-level interface. To use the wrapper APIs in the process model, one needs to include the header file containing the APIs. By default, OPNET Modeler will look for header files in its installation path and in model paths. Therefore, in practice, the wrapper API files can be saved in the model paths to allow OPNET Modeler to pick them up when building the simulation. In this example, the "geo_topo.h" file is saved in the "hlw" folder within a model path. In the HB block of the process model, one can include this wrapper API file path, as shown in Figure 6.2.

Now the user can use all wrapper APIs defined in the "geo_topo.h" file in the process model without problem. It is noted that the model path should be included in the "Model Directories" preference. This preference can be found in OPNET Modeler: go to the "Edit" menu – choose "Preferences" – search for "Model Directories" in the preferences dialog.

```
/* Get next node id from current node's transmitter id */
/* Previous node and current node are connected by link */
/* "recv" is the current node's transmitter module id   */
/* Returns the next node id or -1 if no node connected */
Objid w_get_next_node_from_xmit(Objid xmit)
{
  int count, i;
  Objid link, current_node, next_node;

  FIN(w_get_next_node_from_xmit(xmit));

  if(op_topo_assoc_count(xmit, OPC_TOPO_ASSOC_OUT, \
      OPC_OBJMTYPE_LINK) == 0)
    return -1;
  link = op_topo_assoc(xmit, OPC_TOPO_ASSOC_OUT, \
      OPC_OBJMTYPE_LINK, 0);
  count = op_topo_assoc_count(link, OPC_TOPO_ASSOC_OUT, \
      OPC_OBJMTYPE_NODE);
  current_node = op_topo_parent(op_id_self());
  for(i = 0; i < count; ++i)
  {
    next_node = op_topo_assoc(link, OPC_TOPO_ASSOC_OUT, \
        OPC_OBJMTYPE_NODE, i);
    if(current_node != next_node)
      break;
  }

  FRET(next_node);
}
```

Figure 6.1   Sample wrapper API

```
#include "hlw/geo_topo.h"

#define STREAM (op_intrpt_type () == OPC_INTRPT_STRM)

typedef enum { Destroy, Deliver } pp_status;
static pp_status check_packet (Packet *);
```

Figure 6.2   Code in HB block

6.2   Wrapper APIs provided with the book

In this book we provide several wrapper API packages that can help users to build their models more efficiently. These wrapper APIs are written in C in order to target a wider audience. You can also modify these wrapper APIs to make them more functional. These wrapper API files are included in "hlw" folder and can be downloaded from the publisher's website. You can copy this folder to one of your model paths and include relevant files in the HB block of your process model in order to use these wrapper APIs. All of these wrapper APIs are commented in the source files, but in this section we will explain some of them to help you better understand these wrapper APIs. If a function has "_self" suffix, then this "_self" refers to the current process model's containing module or node. If a wrapper API function is thread-unsafe, this is explicitly commented at the beginning of the function. Otherwise, the function is thread-safe. For thread-unsafe APIs, users should serialize the accessing of shared objects between multiple threads themselves. In parallel simulation, thread-safety should be considered. The topics of parallel simulation and thread-safety are not covered in this version of the book.

6.2.1   Geo_Topo wrapper APIs

In this section, we introduce a wrapper API package called Geo_Topo package, which includes some useful functions to allow users to handle issues of geography, topology, and mobility. For implementations of these APIs in the Geo_Topo package, check the "hlw/geo_topo.h" file which is shipped with this book. The following are descriptions of these functions:

- double w_get_straight_distance(Objid n1, Objid n2) – Computes the straight line distance between two nodes by their object IDs. "n1"and "n2" are object IDs for two nodes. Returns the distance in meters.
- double w_get_straight_distance_by_name(char *name1, char *name2) – Computes the straight line distance between two nodes by name. "name1" and "name2" are full hierarchical names of nodes, like "top.subnet_0.node_0". Returns the distance in meters.
- double w_get_circle_distance(Objid n1, Objid n2) – Computes the great circle distance between two nodes by object IDs. "n1" and "n2" are object IDs for two nodes. Returns the distance in meters.
- double w_get_circle_distance_by_name(char *name1, char *name2) – Computes the great circle distance between two nodes by name. "name1" and "name2" are full hierarchical names of nodes, like "top.subnet_0.node_0". Returns the distance in meters.
- void w_move_node(Objid n, double xpos, double ypos, double altitude) – Change mobile node position by setting the three components of the position of the communications node by object IDs. "n" is the ID of the node to move; "xpos", "ypos", and "altitude" are the three components of the position.
- void w_move_node_self(double xpos, double ypos, double altitude) – Change current mobile node position. "xpos", "ypos", and "altitude" are the three components of the position.
- void w_move_node_by_name(char *name, double xpos, double ypos, double altitude) – Change mobile node position by setting the three components of the position of the

communications node by name. "name" is full hierarchical name of the node to move, like "top.subnet_0.node_0". "xpos", "ypos", and "altitude" are the three components of the position.

- void w_get_node_pos(Objid n, double *xpos, double *ypos, double *altitude) – Get the position of a communication node by object IDs. "n" is the object ID of the node; "xpos", "ypos", and "altitude" are the three components of the position.
- void w_get_node_pos_self(double *xpos, double *ypos, double *altitude) – Get the position of a current communications node. "xpos", "ypos", and "altitude" are the three components of the position.
- Objid w_get_surrounding_module_by_in_strm(Objid module, int strm) – Get the surrounding module object ID from input stream index. "module" is the surrounded module. "strm" is the index of the input stream. Returns the surrounding module ID.
- Objid w_get_surrounding_module_by_in_strm_self(int strm) – Get the ID of current module's surrounding module from the input stream index. "strm" is the index of the input stream. Returns the ID of current module's surrounding module.
- Objid w_get_surrounding_module_by_out_strm(Objid module, int strm) – Get surrounding module ID from the output stream index. "module" is the surrounded module. "strm" is the index of the output stream. Returns the surrounding module object ID.
- Objid w_get_surrounding_module_by_out_strm_self(int strm) – Get the ID of current module's surrounding module from the output stream index. "strm" is the index of the output stream. Returns the ID of current module's surrounding module.
- int w_get_surrounding_modules_in(Objid module, PrgT_List *modules) – Get IDs of surrounding modules that initiate packet streams. "module" is the surrounded module. "modules" receives IDs of surrounding modules initiating packet streams. Returns the number of IDs. This function is thread-unsafe.
- int w_get_surrounding_modules_in_self(PrgT_List *modules) – Get IDs of current module's surrounding modules that initiate packet streams. "modules" receives IDs of surrounding modules initiating packet streams. Returns the number of IDs. This function is thread-unsafe.
- int w_get_surrounding_modules_out(Objid module, PrgT_List *modules) – Get IDs of surrounding modules that receive packet streams. "module" is the surrounded module. "modules" receives IDs of surrounding modules receiving packet streams. Returns the number of IDs. This function is thread-unsafe.
- int w_get_surrounding_modules_out_self(PrgT_List *modules) – Get IDs of the current module's surrounding modules that receive packet streams. "modules" receives IDs of surrounding modules receiving packet streams. Returns the number of IDs. This function is thread-unsafe.
- int w_get_input_indices_from_prev_module(Objid module, PrgT_List *indices) – Get indices of the current module's input streams from the specified previous module. "module" is the previous module initating packet streams. "indices" receives indices of input packet streams. Returns the number of indices. This function is thread-unsafe.
- int w_get_output_indices_from_next_module(Objid module, PrgT_List *indices) – Get indices of the current module's output streams to the specified next module.

"module" is the next module receiving packet streams. "indices" receives indices of output packet streams. Returns the number of indices. This function is thread-unsafe.

- int w_get_surrounding_recvs(Objid module, PrgT_List *recvs) – Get object IDs of surrounding receiver modules. "module" is the surrounded module. "recvs" receives IDs of surrounding receiver modules. Returns the number of IDs. This function is thread-unsafe.

- int w_get_surrounding_recvs_self(PrgT_List *recvs) – Get IDs of the current module's surrounding receiver modules. "recvs" receives IDs of the current module's surrounding receiver modules. Returns the number of IDs. This function is thread-unsafe.

- int w_get_surrounding_xmits(Objid module, PrgT_List *xmits) – Get IDs of surrounding transmitter modules. "module" is the surrounded module. "xmits" receives IDs of surrounding transmitter modules. Returns the number of IDs. This function is thread-unsafe.

- int w_get_surrounding_xmits_self(PrgT_List *xmits) – Get object IDs of the current module's surrounding transmitter modules. "xmits" receives IDs of the current module's surrounding transmitter modules. Returns the number of IDs. This function is thread-unsafe.

- Objid w_get_prev_node_from_recv(Objid recv) – Get the previous node ID from current node's receiver module object ID. Previous node and current node are connected by link. "recv" is the current node's receiver module ID. Returns the previous node ID, or −1 if no node is connected.

- Objid w_get_next_node_from_xmit(Objid xmit) – Get next node ID from the current node's transmitter module object ID. Previous node and current node are connected by link. "recv" is the current node's transmitter module object ID. Returns the next node object ID, or −1 if no node is connected.

- int w_get_connected_nodes_in(PrgT_List *nodes) – Get the current node's connected input nodes by links within the same subnet. "nodes" receives the IDs of the current node's connected input nodes. Returns the number of IDs. This function is thread-unsafe.

- int w_get_connected_nodes_out(PrgT_List *nodes) – Get the current node's connected output nodes by links within the same subnet. "nodes" receives the IDs of current node's connected output nodes. Returns the number of IDs. This function is thread-unsafe.

6.2.2   Routing wrapper APIs

In this section, we will introduce a wrapper API package called Routing package, which includes some useful functions to allow users to handle graph and routing issues. For implementations of these APIs in this package, please check the "hlw/routing.h" file. Some structures are used in these APIs, as defined in Figure 6.3. The following are the descriptions of these functions:

```
typedef struct
{
  /* pointer to the graph */
  PrgT_Graph *graph;
  /* graph state id */
  PrgT_Graph_State_ID state_id;
} W_Graph_Info;

typedef struct
{
  /* pointer to the vertex */
  PrgT_Graph_Vertex *vertex;
  /* object id of the node that is associated
     with the vertex */
  Objid id;
} W_Vertex_Info;

typedef struct
{
  /* source vertex of an edge */
  PrgT_Graph_Vertex *src;
  /* destination vertex of an edge */
  PrgT_Graph_Vertex *dest;
  /* flag that specifies if the edge is duplex or not */
  Boolean duplex;

  /* pointer to the edge */
  PrgT_Graph_Edge *edge;
  /* the weight of the edge */
  double weight;
} W_Edge_Info;
```

Figure 6.3   Structures for wrapper API

- W_Graph_Info w_init_graph(const char *namespace_name, const char *state_name)
  – Initialize and prepare the graph for routing. "namespace_name" and "state_name"
  are names of namespace and state for registering the client state handlers. Returns
  graph info that can be used to manipulate the graph.
- void w_uninit_graph(W_Graph_Info *graph_info) – Uninitialize and destroy the
  graph. "graph_info" refers to the graph info object.
- void w_set_graph_vertices(W_Graph_Info *graph_info, PrgT_List *vertex_info_list)
  – Set vertices for a graph. "graph_info" refers to the graph info object. "ver-
  tex_info_list" contains a list of W_Vertex_Info objects.

- void w_set_graph_edges(W_Graph_Info *graph_info, PrgT_List *edge_info_list) – Set edges for a graph. The "graph_info" refers to the graph info object. "edge_info_list" contains a list of W_Edge_Info objects.
- void w_compute_shortest_path(W_Graph_Info *graph_info, PrgT_List *edge_info_list) – Compute DJK shortest path routing algorithm for a graph. "graph_info" refers to the graph info object. "edge_info_list" contains a list of W_Edge_Info objects.
- PrgT_List * w_get_shortest_path_nodes(W_Graph_Info *graph_info, PrgT_Graph_Vertex *src, PrgT_Graph_Vertex *dest, int index) – "graph_info" refers to the graph info object. "src" is the source vertex. "dest" is the destination vertex. "index" is the index of which shortest path to query in case of multiple shortest paths. Returns a list of object IDs of the nodes along the shortest path. It is the user's responsibility to free this list.

6.2.3   Flow wrapper APIs

In this section, we will introduce a wrapper API package called Flow package, which includes some useful functions to allow users to handle single and multiple connection-oriented communications such as flow and trunk of flows. For implementations of these APIs in this package, please check "hlw/flow.h" file. Some structures are used in these APIs, as defined in Figure 6.4. The following are descriptions of these functions:

```
typedef struct
{
  /* input of the flow */
  Sbhandle begin;
  /* output of the flow */
  Sbhandle end;
  /* weight of the flow */
  double weight;

  /* specify if the flow object is allocated
     via pooled memory */
  Boolean pooled;
} W_Flow;

typedef struct
{
  /* list of flows in the trunk */
  PrgT_List *flows;
  /* weight of the trunk */
  double weight;
} W_Trunk;
```

Figure 6.4   Structures for wrapper API

- W_Flow * w_create_flow(double weight) – Create flow object. "weight" specifies the weight of the flow. Returns the created flow object.
- void w_destroy_flow(W_Flow *flow) – Create flow object. "flow" is the flow to destroy.
- W_Trunk * w_create_trunk(int id, double weight, int flow_num) – Create trunk object. "id" is the ID of the trunk. "weight" specifies the weight of the trunk. "flow_num" is the number of flows to create within the trunk. Returns the created trunk object.
- void w_destroy_trunk(W_Trunk *trunk) – Destroy trunk object. "trunk" is the trunk to destroy.
- void w_trunk_add_flow(W_Trunk *trunk, double weight) – Add a flow to the trunk. "trunk" is the trunk to which the flow is added. "weight" specifies the weight of the flow.
- void w_trunk_remove_flow(W_Trunk *trunk, int index) – Add a flow to the trunk. "trunk" is the trunk from which the flow is removed. "index" is the index of the flow to remove.
- void w_flow_push_packet(W_Flow *flow, Packet *pkt, OpT_Int64 pkt_tag) – Push a packet into the flow. "flow" is the flow into which a packet is pushed. "pkt" is the packet to be pushed into the flow. "pkt_tag" is the user-defined tag associated with the packet.
- OpT_Sar_Size w_flow_size(W_Flow *flow) – Get the total traffic size within a flow. "flow" is the flow. Returns the total traffic size within the flow.
- Packet * w_flow_pop_segment(W_Flow *flow, OpT_Sar_Size segment_size) – Remove a segment from flow with specified size if flow size is greater than 0; otherwise, return nil. "flow" is the flow from which a segment is removed. "segment_size" is the size of the segment. Returns the removed segment.
- void w_flow_push_segment(W_Flow *flow, Packet *segment) – Push a segment into the flow. "flow" is the flow into which a segment is pushed. "segment" is the segment to push.
- int w_flow_packet_count(W_Flow *flow) – Get the total number of packets within a flow. "flow" is the flow. Returns the total number of packets within the flow.
- Packet * w_flow_pop_packet(W_Flow *flow) – Remove a packet from the flow if one exists; otherwise, return nil. "flow" is the flow from which a packet is removed. Returns the removed packet.

6.3 How to write your own wrapper API

This section demonstrates how to create wrapper APIs and use them in the process model. A wrapper API can be any function that does something. However, we would like to show some guidelines on how to write wrapper APIs more efficiently. The following are some guidelines on how to write and compile your own wrapper APIs. However, you are not restricted to following these rules.

```
/* Divide two integers */
/* "a" is the first integer and "b" is the second */
/* Return the result of the division */
int w_div(int a, int b)
{
  int ret;

  FIN(w_div(a, b));

  if(b == 0)
    handle_error();
  else
    ret = a/b;

  FRET(ret);
}
```

Figure 6.5   Sample wrapper API

- You can write your own wrapper APIs in either C or C++. They have their own individual benefits. For efficiency and portability, it is better to encapsulate functionalities into C, which provides opportunities for wider audiences, i.e., so that both C and C++ programmers are able to use them. For simplicity and better encapsulation, C++ is a good option, which can be used to encapsulate functionalities into templates, classes, making them self-contained and easier to use. For how to use C++ in OPNET modeling, refer to Chapter 12.
- It is suggested that you add comments at the beginning of each API function. The comments should explain the functionality this function provides, the input and output parameters, and the return value.
- Wrapper API function naming should follow some simple routines: starting with "w_", the name of the function should reflect the functionality it provides.
- Add necessary error-handling mechanisms to provide more robust APIs and make the code easy to debug.
- It is suggested that you use FIN, FRET, and FOUT macros within the function to facilitate ODB debugging. For details on FIN, FRET, and FOUT macros, refer to Chapter 11.
- There are several different ways of compiling and using your own wrapper API libraries. The first method is to compile the wrapper APIs into dynamic libraries and link these dynamic libraries with simulation kernel executable at runtime. The second method is to compile the wrapper APIs into static libraries and link these static libraries into simulation kernel executable. The third method is simply include the wrapper API files in the simulation model code which will be compiled into simulation kernel executable as a whole. However, in practice the first two methods require more effort to manually manipulate the simulation kernel and wrapper API libraries. For

the third method, the only drawback is the amount of compilation overhead incurred by the included wrapper API files each time you compile your models. However, this overhead is generally negligible.

You can also check the wrapper APIs provided with this book to see how they are implemented. Figure 6.5 shows a simple example that demonstrates how to write a wrapper API by following these guidelines.

7 Modeling with high-level wrapper APIs

This chapter first revisits the cases described in Chapter 5 but utilizing high-level wrapper APIs instead. Then, another case is demonstrated to illustrate how to create connection-oriented communications. To follow this chapter, it is assumed that a reader understands the content covered in Chapter 5 and Chapter 6.

7.1 Revisit of previous case

In this section, Case 6 in Chapter 5 is revisited but this time with wrapper APIs.

Open "pk_switch" process model in Process Editor. Save it as "pk_switch_v2". Now, you can edit "pk_switch_v2" and replace relevant code with wrapper APIs. In SV block, replace the declarations of state variables, as shown in Figure 7.1.

In TV block, replace the declarations of temporary variables, as shown in Figure 7.2.

In HB block, include header files: "routing.h" and "geo_topo.h", as shown in Figure 7.3. These two files include relevant wrapper APIs for performing routing and topology related operations.

In "init" state, replace previous code for building the graph and routing table with the new code that utilizes wrapper APIs, as shown in Figures 7.4 and 7.5.

From Figures 7.4 and 7.5, it is seen that after applying the wrapper APIs, the processing of vertices and edges in the routing graph is performed by dealing with W_Vertex_Info and W_Edge_Info objects. The steps for implementing routing algorithm can be represented by the corresponding wrapper APIs in the following way:

- Initialize graph – w_init_graph().
- Set vertices of the graph – w_set_graph_vertices().
- Set edges of the graph – w_set_graph_edges().
- Compute the shortest path – w_compute_shortest_path().
- Get the shortest path – w_get_shortest_path_nodes().
- Uninitialize the graph if necessary – w_uninit_graph().

In "slot" state, replace the code shown in Figure 7.6 with the code shown in Figure 7.7. The code in both Figure 7.6 and Figure 7.7 is used to get the appropriate packet switching port that leads to the next hop node; however, the code in Figure 7.6 is reduced significantly by utilizing wrapper APIs in "geo_topo" package.

Type	Name	
Evhandle	timeout_handle	
Objid	node_id	
Objid	mod_id	
PrgT_List *	path_nodes_id_list	
W_Graph_Info	graph_info	
PrgT_List *	vertex_info_list	
W_Vertex_Info	vertices_info[7]	
W_Edge_Info	edges_info[8]	
PrgT_List *	edge_info_list	
PrgT_List *	xmits	
PrgT_List *	out_ports	

Figure 7.1 Code in SV block

```
Packet *pk = OPC_NIL;
int header = 0;
int i = 0;
int j = 0;
int count = 0;
int count2 = 0;
Objid next_hop_node_id;
Objid tmp_xmit;
int out_port = 0;
```

Figure 7.2 Code in TV block

```
#include "hlw/routing.h"
#include "hlw/geo_topo.h"

#define STRM (op_intrpt_type() == OPC_INTRPT_STRM)

#define TIMEOUT_INTRPT_CODE 0
#define TIMEOUT ( \
    (op_intrpt_type() == OPC_INTRPT_SELF) && \
    (op_intrpt_code() == TIMEOUT_INTRPT_CODE))

#define SLOT_DURATION 0.2
#define PORT_NUM 5

#define GRAPH_NAMESPACE_NAME "graph_namespace"
#define GRAPH_STATE_NAME "graph_state_name"
```

Figure 7.3 Code in HB block

```
timeout_handle = op_intrpt_schedule_self( \
    op_sim_time() + SLOT_DURATION, \
    TIMEOUT_INTRPT_CODE);

vertices_info[0].id = op_id_from_hierarchical_name( \
    "top.subnet_0.switch_0");
vertices_info[1].id = op_id_from_hierarchical_name( \
    "top.subnet_0.switch_1");
vertices_info[2].id = op_id_from_hierarchical_name( \
    "top.subnet_0.switch_2");
vertices_info[3].id = op_id_from_hierarchical_name( \
    "top.subnet_0.switch_3");
vertices_info[4].id = op_id_from_hierarchical_name( \
    "top.subnet_0.switch_4");
vertices_info[5].id = op_id_from_hierarchical_name( \
    "top.subnet_0.node_0");
vertices_info[6].id = op_id_from_hierarchical_name( \
    "top.subnet_0.node_1");
graph_info = w_init_graph(GRAPH_NAMESPACE_NAME, \
    GRAPH_STATE_NAME);
vertex_info_list = prg_list_create();
for(i = 0; i < 7; ++i)
  prg_list_insert(vertex_info_list, &vertices_info[i], \
      PRGC_LISTPOS_TAIL);
w_set_graph_vertices(&graph_info, vertex_info_list);
```

Figure 7.4   Code in "init" state

Save and compile the "pk_switch_v2" process model. Open chapter5-case6 scenario in Project Editor. Change the process model of "q_0" module in "pk_switch" node model from "pk_switch" to "pk_switch_v2". Now you can run the chapter5-case6 scenario. It works exactly the same as in the case with "pk_switch" as process model. Compared with the code in the "pk_switch" process model, wrapper APIs reduce the programming burden by making simpler code and making the code more readable.

7.2   Creating connection-oriented communications

In this section, we will demonstrate how to model connection-oriented communications via flow wrapper APIs. The flow wrapper API package encapsulates the functions in OPNET Segmentation and Reassembly Package. This package is used to segment a buffer of packets into any size of segments, and it is possible to reassemble them into original packets. This process can be used to model the connection-oriented communication, where at one end source packets are pushed into buffer and the buffer data are

```
edge_info_list = prg_list_create();
edges_info[0].src = vertices_info[0].vertex;
edges_info[0].dest = vertices_info[1].vertex;
edges_info[0].weight = 1;
edges_info[1].src = vertices_info[0].vertex;
edges_info[1].dest = vertices_info[3].vertex;
edges_info[1].weight = 1;
edges_info[2].src = vertices_info[1].vertex;
edges_info[2].dest = vertices_info[2].vertex;
edges_info[2].weight = 1;
edges_info[3].src = vertices_info[1].vertex;
edges_info[3].dest = vertices_info[4].vertex;
edges_info[3].weight = 0.2;
edges_info[4].src = vertices_info[2].vertex;
edges_info[4].dest = vertices_info[4].vertex;
edges_info[4].weight = 1.5;
edges_info[5].src = vertices_info[3].vertex;
edges_info[5].dest = vertices_info[4].vertex;
edges_info[5].weight = 1;
edges_info[6].src = vertices_info[5].vertex;
edges_info[6].dest = vertices_info[3].vertex;
edges_info[6].weight = 1;
edges_info[7].src = vertices_info[6].vertex;
edges_info[7].dest = vertices_info[2].vertex;
edges_info[7].weight = 1;
for(i = 0; i < 8; ++i)
{
  edges_info[i].duplex = TRUE;
  prg_list_insert(edge_info_list, &edges_info[i], \
      PRGC_LISTPOS_TAIL);
}
w_set_graph_edges(&graph_info, edge_info_list);
w_compute_shortest_path(&graph_info, edge_info_list);
path_nodes_id_list = w_get_shortest_path_nodes( \
    &graph_info, vertices_info[5].vertex, \
    vertices_info[6].vertex, 0);

mod_id = op_id_self();
node_id = op_topo_parent(mod_id);
xmits = prg_list_create();
out_ports = prg_list_create();
```

Figure 7.5   Code in "slot" state

```
count = op_topo_assoc_count(mod_id, \
    OPC_TOPO_ASSOC_OUT, OPC_OBJTYPE_STRM);
found = FALSE;
for(i = 0; i < count; ++i)
{
  strm_id = op_topo_assoc(mod_id, \
      OPC_TOPO_ASSOC_OUT, OPC_OBJTYPE_STRM, i);
  tx_id = op_topo_assoc(strm_id, \
      OPC_TOPO_ASSOC_OUT, OPC_OBJTYPE_PTTX, 0);
  if(op_topo_assoc_count(tx_id, OPC_TOPO_ASSOC_OUT,\
      OPC_OBJTYPE_LKDUP) == 0)
    continue;
  link_id = op_topo_assoc(tx_id, OPC_TOPO_ASSOC_OUT, \
      OPC_OBJTYPE_LKDUP, 0);
  count2 = op_topo_assoc_count(link_id, \
      OPC_TOPO_ASSOC_OUT, OPC_OBJTYPE_NODE_FIX);
  for(j = 0; j < count2; ++j)
  {
    tmp_node_id = op_topo_assoc(link_id, \
        OPC_TOPO_ASSOC_OUT, OPC_OBJTYPE_NODE_FIX, j);
    if(tmp_node_id == next_hop_node_id)
    {
      op_ima_obj_attr_get(strm_id, "src stream", \
          &out_port);
      found = TRUE;
      break;
    }
  }
  if(found == TRUE)
    break;
}
```

Figure 7.6   Code in "slot" state

segmented without knowing the boundaries of packets, and at the other end the segments are reassembled into original source packets. In this section, two cases will be modeled: one is single flow, another is a trunk of flows. Here, flow is the same as connection in general, i.e., all segments within the same flow have a common feature, like the same destination address.

7.2.1   Single flow

First, you need to create a node model that is capable of making source packets into flow segments at one end and recovering source packets from flow segments at another end.

```
count = w_get_surrounding_xmits_self(xmits);
for(i = 0; i < count; ++i)
{
  tmp_xmit = *(Objid *)prg_list_access(xmits, i);
  if(w_get_next_node_from_xmit(tmp_xmit) == \
     next_hop_node_id)
  {
    w_get_output_indices_from_next_module( \
        tmp_xmit, out_ports);
    out_port = *(int *)prg_list_access(out_ports, 0);
    break;
  }
}
```

Figure 7.7   Code in "slot" state

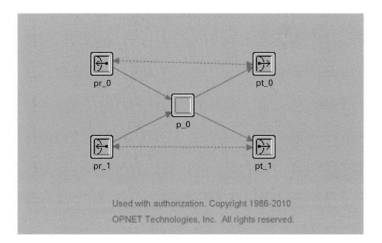

Figure 7.8   Node model

The new node model is shown in Figure 7.8. There are two interfaces in this node model. One interface ("pr_0" and "pt_0") is for receiving and transmitting source packets and another interface ("pr_1" and "pt_1") is for receiving and transmitting flow segments. "p_0" processor module is for handling the traffic from or to these interfaces.

Save the node model as "flow_handler". Next, you can create a new process model that will be used by "p_0" module in "flow_handler" node. The state transition diagram of this process model is shown in Figure 7.9.

In SV block, add declarations of state variables like flow object, as shown in Figure 7.10.

In TV block, add declarations of temporary variables, as shown in Figure 7.11.

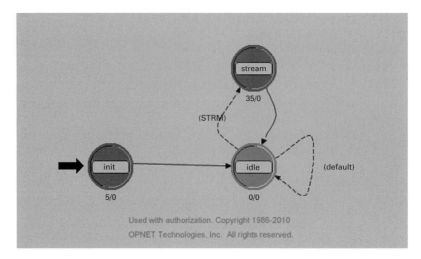

Figure 7.9   State transition diagram

Type	Name	
W_Flow *	flow	
OpT_Int64	pk_num	

Figure 7.10   Code in SV block

```
Packet *pk = OPC_NIL;
Packet *src_pk = OPC_NIL;
Packet *segment = OPC_NIL;
Packet *flow_segment = OPC_NIL;
int in_port;
char pk_fmt[128] = "";
OpT_Sar_Size size;
int i;
int count;
```

Figure 7.11   Code in TV block

In HB block, include the header file for flow wrapper package, as shown in Figure 7.12. In HB block, segment size is also defined, which is independent of source packet sizes.

In "init" state, add code to create a flow object, as shown in Figure 7.13.

In "stream" state, add code to handle traffic from two interfaces, as shown in Figure 7.14: if the format of the received packet is not "flow_segment", it is a source

```
#include "hlw/flow.h"

#define STRM (op_intrpt_type() == OPC_INTRPT_STRM)
#define SEGMENT_SIZE 128
```

Figure 7.12   Code in HB block

```
flow = w_create_flow(0);
pk_num = 0;
```

Figure 7.13   Code in "init" state

packet and it should be pushed into flow, and data is sent out of the flow as segments; if the format of the received packet is "flow_segment", it is a flow segment and it should be pushed into flow, and source packets are recovered and removed from the flow.

In Figure 7.14, the formatted packet "flow_segment" should be created in Packet Format Editor. There are two fields in "flow_segment" format. "flow_id" is an integer type, and "segment" is a packet type with "inherited" size. "flow_id" stores the identifier of the flow that this segment belongs to. "segment" stores the actual segment data packet. It is shown in Figure 7.15.

Save this process model as "flow_handler". From the "Interfaces" menu of Process Editor, choose "Process Interfaces". Set "begsim intrpt" and "endsim intrpt" attributes to "enabled", as shown in Figure 7.16.

Save and compile "flow_handler" process model. set the "process model" attribute of "p_0" module in "flow_handler" node to "flow_handler".

Next, you can create a simulation scenario for this model. Create a new project and scenario with project name "chapter7" and scenario name "case1". Create a network topology as shown in Figure 7.17. The model of "flow_handler_0" and "flow_handler_1" nodes is "flow_handler". The model of "source_node_0" and "source_node_1" nodes is "basic_source", which was created in Chapter 5. The model of the links between these nodes is "basic_link", which was created in Chapter 5.

In Project Editor, from the "DES" menu, check "Record Packet Flow 2D Animation For Subnet". Press "Ctrl+R" and "Alt+R" consecutively to start simulation. After simulation finishes, in Project Editor, from the "DES" menu choose "Play 2D Animation". In Animation Viewer, it is seen that packets flow faster between "flow_handler_0" and "flow_handler_1" nodes than between "source_node_0" and "flow_handler_0" nodes, and faster than between "flow_handler_1" and "source_node_1" nodes. This is because the packets flowing between "flow_handler_0" and "flow_handler_1" nodes are actually flow segments that have smaller sizes than the original source packets.

```
in_port = op_intrpt_strm ();
pk = op_pk_get (in_port);
op_pk_format (pk, pk_fmt);
if(strcmp(pk_fmt, "flow_segment") != 0)
{
  // If packet format is not flow segment,
  // add this packet to outgoing flow, get
  // segments from the flow and send them out
  w_flow_push_packet (flow, pk, pk_num++);
  size = w_flow_size (flow);
  for(; size > 0; size -= SEGMENT_SIZE)
  {
    segment = w_flow_pop_segment (flow, SEGMENT_SIZE);
    flow_segment = op_pk_create_fmt ("flow_segment");
    op_pk_nfd_set (flow_segment, "flow_id", 0);
    op_pk_nfd_set (flow_segment, "segment", segment);
    if(in_port == 0)
      op_pk_send (flow_segment, 1);
    else
      op_pk_send (flow_segment, 0);
  }
}
else
{
  // If packet format is flow segment,
  // add this segment into incoming flow,
  // recover original packets from the flow
  // and send recovered packets to destination
  op_pk_nfd_get (pk, "segment", &segment);
  w_flow_push_segment (flow, segment);
  count = w_flow_packet_count (flow);
  for(i = 0; i < count; ++i)
  {
    src_pk = w_flow_pop_packet (flow);
    if(in_port == 0)
      op_pk_send (src_pk, 1);
    else
      op_pk_send (src_pk, 0);
  }
  op_pk_destroy (pk);
}
```

Figure 7.14   Code in "stream" state

7.2.2 Trunk of flows

Now, you can modify "flow_handler" process model so that it can model the behaviors of a number of flows. In wrapper flow package, a number of flows can be modeled via the W_Trunk object.

Open "flow_handler" process model in Process Editor. In SV block, declare state variables like trunk object, as shown in Figure 7.18.

In TV block, declare temporary variables as shown in Figure 7.19.

In "init" state, add code to create a trunk object, as shown in Figure 7.20.

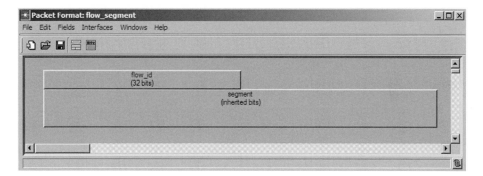

Figure 7.15   Packet format

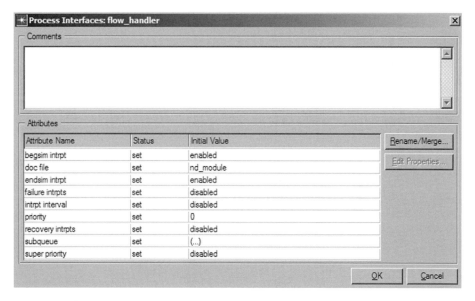

Figure 7.16   Process interfaces

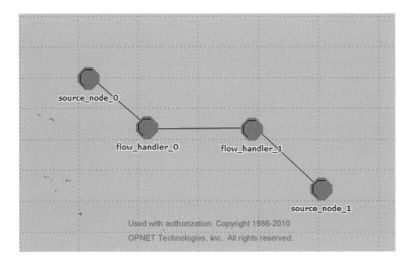

Figure 7.17   Network topology

Type	Name	
W_Trunk *	trunk	
OpT_Int64	pk_num	

Figure 7.18   Code in SV block

```
W_Flow *flow;
Packet *pk = OPC_NIL;
Packet *src_pk = OPC_NIL;
Packet *segment = OPC_NIL;
Packet *flow_segment = OPC_NIL;
int in_port;
char pk_fmt[128] = "";
OpT_Sar_Size size;
int i;
int count;
int flow_id;
```

Figure 7.19   Code in TV block

```
trunk = w_create_trunk(0, 0, FLOW_NUM);
pk_num = 0;
```

Figure 7.20   Code in "init" state

```
in_port = op_intrpt_strm();
pk = op_pk_get(in_port);
op_pk_format(pk, pk_fmt);
if(strcmp(pk_fmt, "flow_segment") != 0)
{
  // If packet format is not flow segment,
  // add this packet to outgoing flow identified by
  // flow_id, get segments from the flow, set "flow_id"
  // attributes for these segments and send them out
  flow_id = pk_num % FLOW_NUM;
  flow = (W_Flow *)prg_list_access(trunk->flows, flow_id);
  w_flow_push_packet(flow, pk, pk_num++);
  size = w_flow_size(flow);
  for(; size > 0; size -= SEGMENT_SIZE)
  {
    segment = w_flow_pop_segment(flow, SEGMENT_SIZE);
    flow_segment = op_pk_create_fmt("flow_segment");
    op_pk_nfd_set(flow_segment, "flow_id", flow_id);
    op_pk_nfd_set(flow_segment, "segment", segment);
    if(in_port == 0)
      op_pk_send(flow_segment, 1);
    else
      op_pk_send(flow_segment, 0);
  }
}
else
{
  // If packet format is flow segment, get "flow_id"
  // attribute value which is used to identify the
  // incoming flow this segment belongs to, add this
  // segment into this incoming flow, recover original
  // packets from the flow and send them to destination
  op_pk_nfd_get(pk, "flow_id", &flow_id);
  op_pk_nfd_get(pk, "segment", &segment);
  flow = (W_Flow *)prg_list_access(trunk->flows, flow_id);
  w_flow_push_segment(flow, segment);
  count = w_flow_packet_count(flow);
  for(i = 0; i < count; ++i)
  {
    src_pk = w_flow_pop_packet(flow);
    if(in_port == 0)
      op_pk_send(src_pk, 1);
    else
      op_pk_send(src_pk, 0);
  }
  op_pk_destroy(pk);
}
```

Figure 7.21   Code in "stream" state

In "stream" state, add code to evenly allocate received source packets to all the flows in the trunk and recover original source packets from flow segments in the corresponding flow. This is shown in Figure 7.21.

Save and compile this process model. Now you can run the simulation to model many flows. You can modify this example to model hierarchical and QoS (quality of service)-based flows. With the flow wrapper API package, you can easily model and manage connection-oriented communications.

Part III

Modeling and Modifying Standard Networks and Protocols

8 Modeling wired networks with standard models

This chapter shows how to construct wired networks with standard OPNET models in an evolutionary style. Differently from Chapters 5–7, where custom OPNET models are developed, in this chapter standard OPNET models are discussed. These standard models are pre-created and shipped with the OPNET Modeler. Standard OPNET models include TCP, IP, IPV6, ATM, MPLS, OSPF, TDMA, WiMAX, ZigBee, etc. To follow this chapter, it is preferable to know the basic operations of OPNET Modeler, though not necessary.

The process for simulating standard models is similar to that for simulating custom models as demonstrated in Chapter 5. The following are the steps of simulating standard models:

- Create a project via Project Editor.
- Create a scenario within this project.
- Create a network topology for this scenario by placing the standard models on the Project Editor.
- Verify the link connectivity of this network.
- Select the statistics in which you are interested.
- Run the simulation for this scenario.
- View, compare, and analyze statistic results.
- If necessary, export statistic data to a spreadsheet for further processing.

In the following sections, case studies on how to model standard wired networks are demonstrated.

8.1 Client/server structure

In this section, a simple client/server structure will be created and modeled. The client and server are computers that are connected by links.

8.1.1 Creating a network model

To create a client/server network model, go through the following steps:

- Create a new project and scenario with project name "chapter8" and scenario name "case1". In Project Editor, press "Open Object Palette" toolbar button. On the right side of the palette dialog, choose "Subnet" object and place this object on the Project Editor, as shown in Figure 8.1.
- Double click the "Subnet" object to go inside the subnet.
- Place "ethernet_wkstn" (client) and "ethernet_server" (server) objects in the subnet.
- Place "Task Config", "Application Config", and "Profile Config" objects onto subnet.
- Use a link to connect the client and server. The link model is "10BaseT". This structure is shown in Figure 8.2.

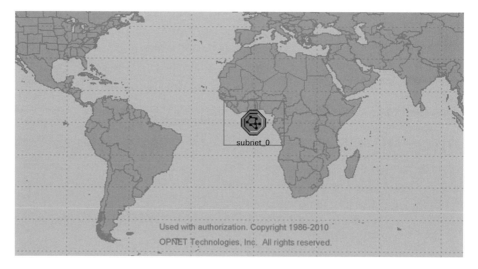

Figure 8.1   Network model

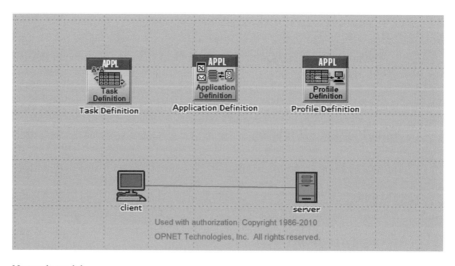

Figure 8.2   Network model

Now you can verify link connectivity. From "Topology", choose "Verify Links..." to verify whether nodes and links are correctly connected. If there are connectivity errors, a red cross will be shown on the link. To resolve the connectivity problem, right-click link object, select "Edit Attributes (Advanced)" to modify "transmitter" and "receiver" attributes appropriately. Since a link connects two nodes, the values of transmitters and receivers should be set to corresponding transceiver objects for connected nodes. After configuration, verify link connectivity again.

Q8.1 When running a simulation, why are there "Recoverable Errors" saying "Packets with packet format (???) are not supported by link or transceiver channel"?

You see this error because the links and transceivers (transmitters and receivers) have incompatible configurations. For example, a duplex link object is connected to a node object's transceiver object. The link object's data rate is "1024 bps" and supported packet format is "ip_dgrm_v4". The transceiver object's data rate is "1024 bps" and supported packet format is "ethernet_v2". The data rate attributes of link object and transceiver object are the same. However, the supported packet formats of link and transceiver do not match each other. Then, if you run simulation, there will be an error report about this mismatch. Therefore, before starting simulation, you should verify link connectivity and clear possible errors. If there are connectivity errors, you can change the corresponding link and transceiver's "data rate" and/or "packet formats" attributes appropriately to make them match each other. For some custom node models, if you are not sure what data rate this node's transceivers should support at the beginning, you can set the value of the "data rate" attribute for the transceiver to "unspecified", which means it supports any data rate. By setting this special value, links of any data rate can be connected to this node's transceivers without a problem.

8.1.2 Task, application, and profile configurations

In this scenario, there are a client and a server connected to each other. However, at this moment there is no traffic running between the client and server. The client and server are models of computer. The internal node model structure is shown in Figure 8.3.

The node structure shows a typical protocol stack for a computer. From this protocol stack, you can see that the root is "application" module. The source of traffic is also "application" module. In the real world, the application in computer can be web, email, database, video/audio, etc, and these applications are sources of traffic flowing between computers. In the simulation model, applications have the same definitions. Therefore, in order to generate traffic for this simulation model, you need to define model applications. There are three objects that can be used to define model applications in OPNET: Task Config, Application Config, and Profile Config. These objects can be found in the Object Palette. The relationship between these configuration objects is: Task Config → Application Config → Profile Config. "Task Config" defines the fundamental traffic features for tasks. These features include task phases, source/destination, request/response, request size, interarrival times and timeout, etc. With "Task Config", you can define custom tasks. The custom task can be referenced in "Application Config" object to define an

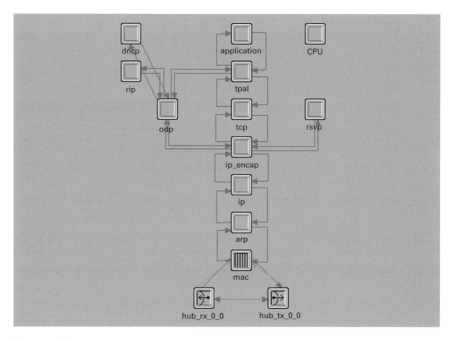

Figure 8.3   Node model

application. However, "Application Config" also contains some pre-defined application types such as "Database", "Email", "Ftp", "Http", "Voice" and "Video Conferencing". If your application falls into these predefined application types, you do not need to define tasks for the application. In "Profile Config", a profile can be defined by incorporating several applications. These applications can be operated in serial or simultaneous mode and each application's duration and repeatability can be configured as well. Next, you can define a profile to be used by client and server to generate traffic. In this profile, there are three predefined applications: Database, Http, and Video Conferencing. These applications should be set in the "Application Config" object. To set these applications, right click "Application Config" object, select "Edit Attributes (Advanced)", edit "Application Definitions" attribute. In the Application Definitions table, add a row called "My Database". In the Description table of "My Database" application, set the value of the "Database" attribute to "Medium Load" as shown in Figure 8.4. This defines "My Database" as a medium-load database application. In the same way, add "My Http" and "My Video Conferencing" applications as well. For "My Http" application, set the "Http" attribute value to "Heavy Browsing". For "My Video Conferencing" application, set the "Video Conferencing" attribute value to "Low Resolution Video".

In this model, since there is no custom application, the "Task Config" object is not used. However, you can define tasks in the "Task Config" object and set these tasks to custom application in "Application Config" in the same way as for a predefined application.

Now you can define the profile via "Profile Config" object. To define the profile, right click "Profile Config" object, select "Edit Attributes (Advanced)", edit "Profile

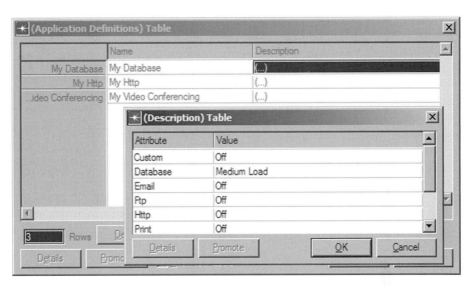

Figure 8.4   Application definitions

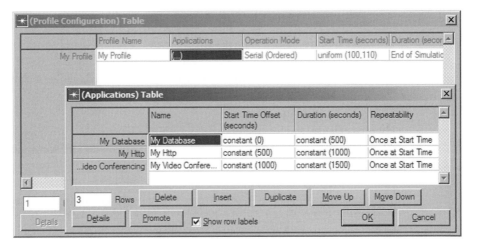

Figure 8.5   Profile configuration

Configuration" attribute. In the Profile Configuration table, add a new row of profile called "My Profile". In this example, set "Operation Mode" to "Serial (Ordered)". Then, you can add three application rows to the Application table of "My Profile". Each row refers to an application set in "Application Config". The configuration of "My Profile" is shown in Figure 8.5.

Next, you can set this profile in client and server nodes so that application traffic can be generated. For client and server nodes, there are three node attributes often used for interacting with application traffic: "Application: Supported Profiles", "Application: Supported Services", and "Application: Destination Preferences". If a computer acts as application traffic originator, a profile should be set to the "Application: Supported Profiles" attribute. If a computer acts as a server, then the "Application: Supported Services" attribute should be configured to support relevant applications defined in the "Application Config" object. However, in some cases, a computer can act as both application traffic originator and server at the same time. For these cases, both "Application: Supported Profiles" and "Application: Supported Services" attributes should be set appropriately. You can specify traffic destination by setting the "Application: Destination Preferences" attribute. However, if the value of the "Application: Destination Preferences" attribute is set to "None", a random destination will be chosen from among the destination nodes that support the application of interest. Selection weight specified in the "Application: Supported Services" attribute on the destination will determine the probability with which the destination will get chosen. You can go through the following steps to configure client and server nodes:

- Right click "client" object, select "Edit Attributes (Advanced)", edit "Application: Supported Profiles" attribute and add a "My Profile" row as shown in Figure 8.6.
- Right click "client" object, select "Edit Attributes (Advanced)", set "Application: Destination Preferences" attribute to "None". Alternatively, in this example, you can set destination to "server" node explicitly as shown in Figure 8.7.
- Right click "server" object, select "Edit Attributes (Advanced)", set "Application: Supported Services" attribute to "All".
- Right click "server" object, select "Edit Attributes (Advanced)", set "Application: Destination Preferences" attribute to "None". Alternatively, in this example, you can set destination to "client" node explicitly.

In this example, we set the profile only for client node but not server node. This is because we only want client node to act as traffic originator and server node to act as

Figure 8.6   Application: supported profiles

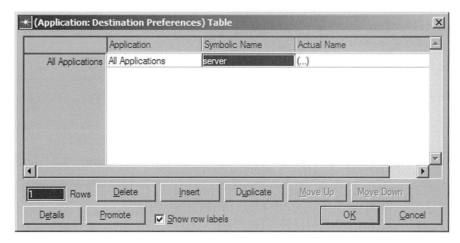

Figure 8.7   Application: destination preferences

service provider. If you want these two nodes to act as both client and server at the same time, then you should set the profile for both nodes.

Until now, the model has been configured to run application traffic. However, before starting simulation, you need to choose the statistics of interest.

8.1.3   Choosing and viewing statistic results

For standard models, there are many predefined statistics. In this example, you can choose three statistics related to the applications for client node as follows:

- Right click "client" object, select "Choose Individual DES Statistics" to show statistics dialog.
- From "Node Statistics", choose "Client DB" application statistics.
- From "Node Statistics", choose "Client Http" application statistics.
- From "Node Statistics", choose "Video Conferencing" application statistics.

Q8.2   What are the differences between "Choose Individual DES Statistics" and "Choose Statistics (Advanced)"?

"Choose Individual DES Statistics" allows you to simply choose the statistics of interest for any object. "Choose Statistics (Advanced)" will invoke "Probe Editor", which can be used to choose statistics to probe for any object, edit probe attributes, change probe capture mode, create new statistic probe, and so on.

In Project Editor, you can start simulation from the "DES" menu by choosing "Run Discrete Event Simulation". After completion of simulation, right click "client" object and select "View Results". In the "Object Statistics" list, choose "Traffic Sent (bytes/sec)" statistic for "Client DB", "Client Http", and "Video Conferencing" applications. The results are shown in Figure 8.8.

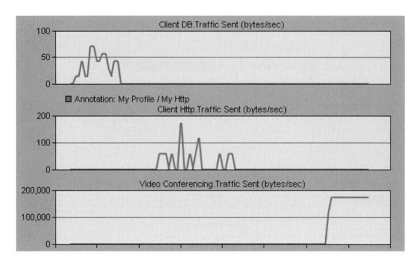

Figure 8.8   Statistic results

From Figure 8.8, you can see that three applications are scheduled in serial order as configured in "My Profile". You can also choose other statistics of interest and run simulation again to view results.

8.2   Local area network

In this section, you can build a Local Area Network (LAN) model based on the client/server model made in the last section. This scenario demonstrates how to build a switched local area network model. You can go through the following steps to build this LAN model:

- In Project Editor, from "Scenarios" menu, choose "Duplicate Scenario..." to create a new scenario called "case2".
- Copy/paste client node four times to reproduce four client nodes.
- In Object Palette, find "ethernet16_switch_adv" model and place it on Project Editor.
- Connect all client and server nodes to the switch with "10BaseT" links as shown in Figure 8.9.

From "Topology", choose "Verify Links..." to verify link connectivity. If there is no problem, you can choose statistics of interest and start simulation. In this case, all client nodes will generate application traffic that is switched to server node, and server node will provide supported application services for all client nodes.

8.3   Wide area IP network

First, create a new scenario called "case3". There are two subnets in this scenario. The traffic in these subnets is routed to an IP cloud object via gateway routers. The topmost network topology is shown in Figure 8.10.

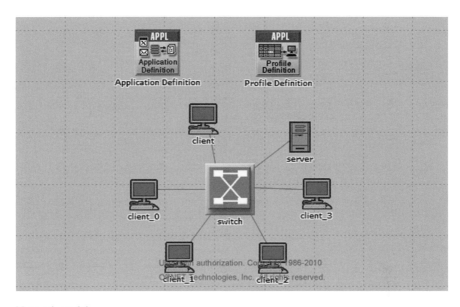

Figure 8.9 Network model

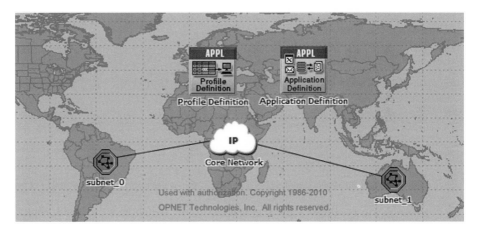

Figure 8.10 Network model

The "Application Definition" and "Profile Definition" objects should be configured in a similar way to that in the client/server scenario. The Core Network is an "ip32_cloud" model object. Two subnets and Core Network are connected by links of "PPP_DS3" model. In "subnet_0", two client nodes are connected to a switch which is connected to a gateway router of the "ethernet4_slip4_gtwy_adv" model. The connection links are objects of the "10Base" model. The client nodes' "Application: Supported Profile" attribute should be configured in a similar way to that in the client/server scenario. The topology of "subnet_0" is shown in Figure 8.11.

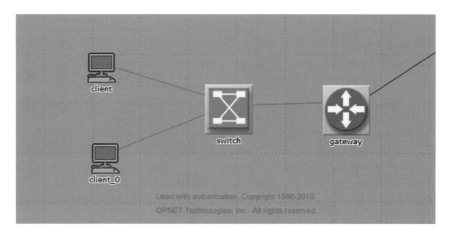

Figure 8.11   Network model

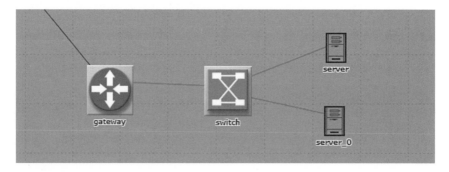

Figure 8.12   Network model

In "subnet_1", two server nodes are connected to a switch which is also connected to a gateway router. The server nodes' "Application: Supported Services" attribute should be configured appropriately. The topology of "subnet_1" is shown in Figure 8.12.

From "Topology", choose "Verify Links..." to verify link connectivity. If there is no problem, you can choose statistics of interest and start simulation.

Until now, you've created a client/server model, a local area network model, and a wide area IP network model. However, by following similar techniques, you can create other models as well.

8.4   Automatic network deployment

In previous sections, you manually configured the network nodes, links, and topologies. However, OPNET Modeler provides a rapid configuration tool that allows you to deploy network topology in an automatic way. The rapid configuration tool not only deploys topology, but also allows you to choose node model and link model to deploy via a

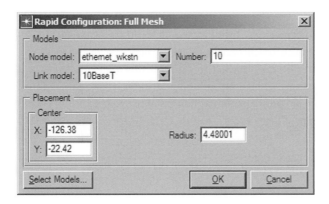

Figure 8.13 Rapid configuration

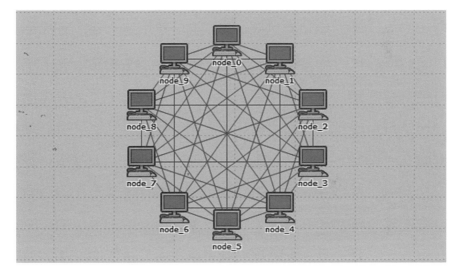

Figure 8.14 Network model

deployment wizard. It is suitable for deploying a network that has typical or regular topology. In Project Editor, from the "Topology" menu, choose "Rapid Configuration" to show the configuration wizard as shown in Figure 8.13.

Figure 8.14 shows a full-meshed network deployed via the Rapid Configuration wizard.

Rapid configuration allows you to quickly deploy baseline network. However, you can make more detailed and customized configurations based on the deployed network.

8.5 Summary

In this chapter, we demonstrated the basics of how to model networks with standard OPNET models including standard node model, link model, task/application/profile model, etc. These models implement standard protocols and algorithms. We also showed how to apply practical applications and the profile of applications to workstations and servers to generate appropriate traffic based on these applications. However, there are other ways of generating traffic between standard nodes, such as generating traffic based on the traffic characteristics (packet interarrival times and packet size distributions) rather than applications, generating self-similar traffic (Park and Willinger 2000) and even generating hybrid traffic (some explicit discrete event traffic and some background analytical traffic), etc. For more information on traffic generation, check Chapter 13.

9 Modeling wireless networks with standard models

This chapter explains the concepts and techniques of building wireless networks based on standard OPNET models. It is essential to understand the concepts described in Chapter 8 when reading this chapter, and a basic knowledge of wireless technology is required.

9.1 Basics of wireless modeling

The processes for modeling wireless networks are similar to those for modeling wired networks in OPNET Modeler. However, there are some differences. For wireless networks, nodes are connected via invisible radio links instead of cable links. Unlike fixed cable links, a radio link can be influenced by interference, antenna pattern, and movement of mobile nodes. Therefore, it requires simulation to dynamically compute radio link connectivity, propagation delay, and power levels.

In Node Editor, there are three modules used for modeling wireless nodes: Radio transmitter, Radio receiver and Antenna, as shown in Figure 9.1.

In Figure 9.1, the module marked no. 1 is the radio receiver module, the module marked no. 2 is the radio transmitter module, and the module marked no. 3 is the antenna module. Radio transceiver modules allow packets to be sent or received via radio links. The antenna module can be used to exchange packets with other nodes when antenna directionality or gain needs to be modeled. Antenna module is associated with Radio transmitter and Receiver modules by means of packet streams as shown in Figure 9.1. Antenna location in three-dimensional space can be determined by Antenna module's "latitude", "longitude", and "altitude" attributes. Antenna module can be pointed by setting Antenna module's "pointing ref.phi" and "pointing ref.theta" attributes. Antenna pattern maps gain to all directions in a three-dimensional object whose shape indicates the relative magnitudes of gain in each direction. Antenna pattern can be designed via Antenna Pattern Editor, which can be invoked from "File" menu – "New..." – "Antenna Pattern". An antenna pattern model is shown in Figure 9.2.

Figure 9.2 shows the antenna pattern model representing an isotropic pattern which radiates power equally in all directions, and its gain is equal to 0 dB in all directions. You can design other custom antenna patterns within Antenna Pattern Editor by adjusting power levels appropriately in all directions. An antenna pattern model can be associated with an antenna module by specifying the "pattern" attribute of the Antenna module. Antenna module's behaviour will be based on the associated antenna pattern.

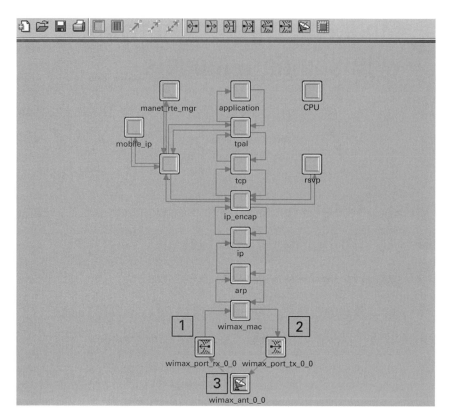

Figure 9.1   Node model

9.2   Wireless local area networks (WLANs)

Modeling WLANs (Bing 2002) is similar to modeling LANs. The main difference is that clients and server communicate via radio links in a WLAN instead of visible wired links. You can create two network scenarios in this section: one for communication within a WLAN and another for communication between WLANs.

9.2.1   Communication within WLANs

First, create a project called "chapter9" and scenario called "case1". The network domain topology of a WLAN is shown in Figure 9.3. The model of client nodes is mobile "wlan_wkstn_adv" and the model of the server node is "wlan_server_adv".

Next, define an HTTP application called "My Http" via the Application Config object and set the "Http" attribute to "Heavy Browsing", as shown in Figure 9.4.

Define a profile called "My Profile" via the Profile Config object and add "My Http" application to Applications table as shown in Figure 9.5.

For all client nodes, set "Application: Supported Profiles" to "My Profile". For server node, set "Application: Supported Services" to "All". Before starting simulation, you

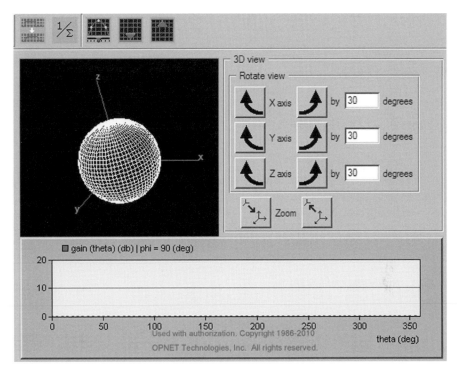

Figure 9.2   Antenna pattern model

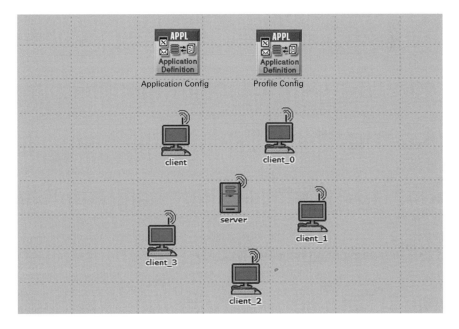

Figure 9.3   Network model

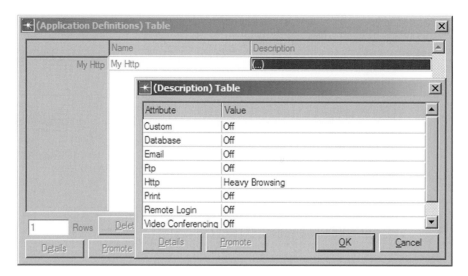

Figure 9.4   Application Definitions

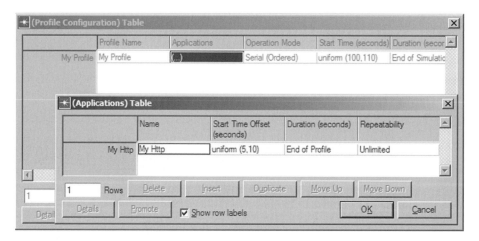

Figure 9.5   Profile configuration

should choose statistics of interest. In this example, you can choose "Global Statistics – HTTP" statistics. After simulation completes, you can view statistic results such as "Page Response Time" and "Traffic Sent", as shown in Figure 9.6.

9.3   Communication between WLANs

For "case2" scenario, you can copy "case1" scenario by choosing "scenario" menu – "Duplicate Scenario...". Add a node object of "wlan2_router_adv" model type. This node acts as a router responsible for communications between two WLANs. The network topology is shown in Figure 9.7.

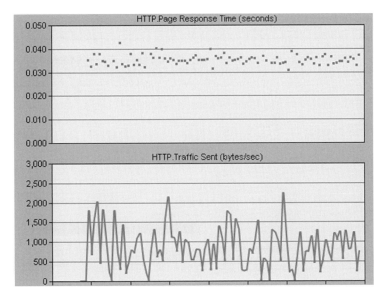

Figure 9.6 Statistic results

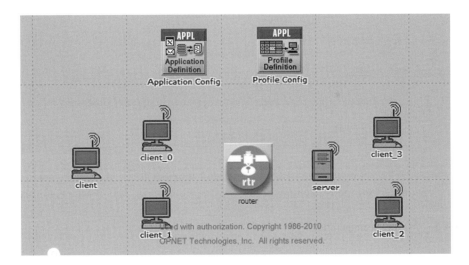

Figure 9.7 Network model

A WLAN is defined by the basic service set (BSS), which is a set of wireless nodes that can communicate with each other. In order to group wireless nodes into two WLANs, we should explicitly configure these nodes. Select "client", "client_0", and "client_1", right click the selection, choose "Edit Attributes (Advanced)" to show the Attributes table. Set the "BSS Identifier" attribute to "0" from the "Wireless LAN" – "Wireless LAN Parameters" table, as shown in Figure 9.8. Setting this attribute will explicitly make "client", "client_0", and "client_1" wireless nodes in BSS 0 wireless LAN. In the same way, set "client_2", "client_3", and "server" nodes' "BSS Identifier" attribute

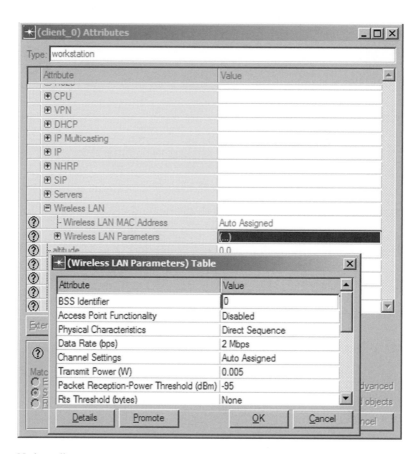

Figure 9.8   Node attributes

to "1". Now there are BSS 0 and BSS 1 wireless LANs in the network. To make BSS 0 network nodes communicate with BSS 1 network nodes, we need to configure the "router" object. The "router" object has two wireless LAN interfaces. We should make one interface interact with BSS 0 network and another interact with BSS 1 network. Right click the "router" object, choose "Edit Attributes (Advanced)" to show Attributes table. Set "Wireless LAN" – "Wireless LAN Parameters (IF0 P0)" – "BSS Identifier" to "0" and set "Wireless LAN" – "Wireless LAN Parameters (IF1 P0)" – "BSS Identifier" to "1".

You can run two simulations and compare results for the two simulations. One simulation is with "router" object disabled and another is with "router" object enabled. To disable an object, select objects, then press the "Fail Selected Objects" tool button. To enable an object, select objects, then press the "Recover Selected Objects" tool button. The statistic we are interested is "Traffic Received" for "server" object. Right click "server" object and select "Choose Individual DES Statistics". In "Choose Results" dialog, choose "Node Statistics" – "Server Http" – "Traffic Received (bytes/sec)". Next, run two simulations. Figure 9.9 shows "Traffic Received (bytes/sec)" statistics for two simulations. The statistic results are averaged.

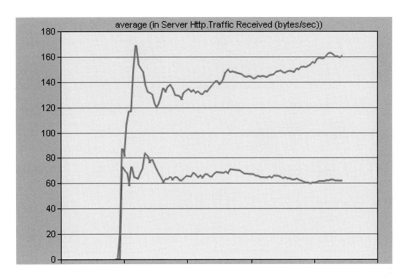

average (in Server Http.Traffic Received (bytes/sec))

Figure 9.9   Statistic results

In Figure 9.9, the lower trace is the traffic "server" received when "router" is disabled and the upper trace is when "router" is enabled. The compared results can be interpreted as follows: when "router" is disabled, "server" received traffic only from "client_2" and "client_3" nodes, since these three wireless nodes are in the same WLAN, i.e., BSS 1; when "router" is enabled, "router" can route traffic from BSS 0 to BSS 1, so "server" can receive traffic sent from "client", "client_1", and "client_2" nodes in WLAN BSS 0 as well.

9.4   Wireless mobile networks

In wireless networks, since wireless nodes may change their positions, modeling node movement is necessary. In this section, three techniques used to model mobile networks will be described: movement via trajectories, facilities for random mobility, and movement via programming interfaces.

9.4.1   Movement via trajectories

There are two types of trajectory in OPNET Modeler: segment-based and vector-based trajectories. Segment-based trajectories define movement path with a series of segments separated by pre-defined points. Vector-based trajectories define movement via bearing, ground speed, and ascent rate attributes of a mobile node.

Segment-based trajectory

To demonstrate segment-based trajectory, you need a new scenario. The following procedure creates a prototype scenario "case3" based on scenario "case1".

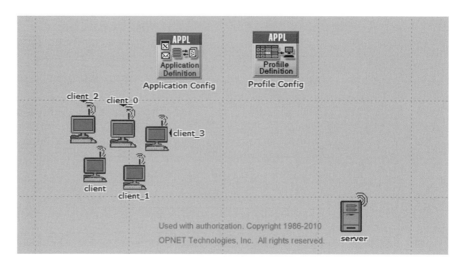

Figure 9.10   Network model

- Open "chapter9" project. Switch to "case1" scenario from the menu "Scenarios" – "Switch to Scenario" – "case1".
- Copy scenario "case1" from menu "Scenarios" – "Duplicate Scenario...". Name the duplicated scenario "case3".
- Change the positions of these wireless nodes so that the network looks like that shown in Figure 9.10.

Next, you can define a segment-based trajectory. A segment-based trajectory is composed of a series of path segments. Each path segment can have individual features. There are two segment-based trajectory types: fixed-interval and variable-interval. For fixed-interval trajectory, a mobile node takes the same amount of time to traverse every path segment. For variable-interval trajectory, each point along the trajectory has its own specified altitude, wait time, segment traversal time, and orientation. The wait time makes a mobile node pause at each segment point before traversing the next segment. You can go through the following steps to create a segment-based trajectory object with fixed interval. It is similar to creating a segment-based trajectory with variable interval. The main difference is that when defining each path segment, the "Segment Information" dialog will show up to ask you to specify parameters for this defined segment, since each segment can have individual parameters in variable-interval trajectory.

- From the menu "Topology", choose "Define Trajectory..." to show "Define Trajectory" dialog.
- Make the Trajectory name "trajectory1".
- For Trajectory type, choose "Fixed interval".
- Set Time step to "0h30m0s", i.e., 30 minutes for traversing each trajectory segment.
- Check the "Coordinates are relative to object's position" checkbox. This is to make sure the trajectory's initial position can follow the node object's initial position on the network.

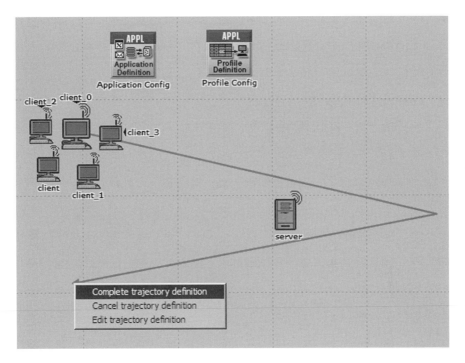

Figure 9.11 Network model

- Press the "Define Path" button. Work with the mouse to define the segments of the path on the network. Define two segments to form a trajectory path as shown in Figure 9.11.
- Right click mouse at workspace, choose "Complete trajectory definition" to finish defining the trajectory.
- Select all clients, right click selected objects, choose "Edit Attributes (Advanced)" to show the "Attributes" dialog, set the "Trajectory" attribute to "trajectory1". Now, all clients have "trajectory1" as their movement trajectories, as shown in Figure 9.12.

Now you can choose the statistics in which we are interested and run simulation. To view node movement in OPNET Animation Viewer, in Project Editor, you should check the menu "DES" – "Record Node Movement 2D Animation for Subnet" before starting simulation. After simulation completes, choose the "DES" menu – "Play 2D Animation" to play animation in Animation Viewer.

Vector-based trajectory

In contrast to segment-based trajectory, there is no explicit end point for vector-based trajectory. Vector-based trajectory follows the circular path around the Earth. The path is determined by the bearing, ground speed, and ascent rate attributes of the mobile node. First, you can create a new scenario "case4" by duplicating scenario "case3". You can go through the following steps to make clients have vector-based trajectories:

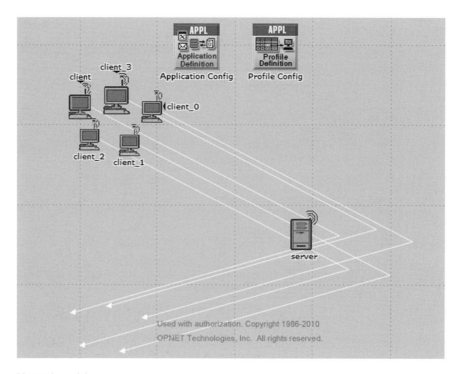

Figure 9.12   Network model

- Select all clients, right click selected objects, choose "Edit Attributes (Advanced)" to show Attributes dialog, set "Trajectory" attribute to "VECTOR".
- Set "Bearing" attribute to "120". This attribute is to specify bearing off magnetic north: 0, North; 90, East; 180, South; 270, West.
- Set "ground speed" to "1 meter/sec". This attribute is to specify the initial speed relative to the ground.
- Set "ascent rate" to "1 meter/sec". This attribute is to specify the vertical speed.
- Press the OK button to confirm configurations. Figure 9.13 shows scenario "case4" after configuration.

Now you can choose the statistics of interest and run the simulation. Unlike segment-based trajectory, which uses multiple segments to form a trajectory path, vector-based trajectory relies on the circle around the Earth, so the trajectory path is determined once the bearing, ground speed, and ascent rate are determined.

9.4.2   Facilities for random mobility

OPNET Modeler provides a "Mobility Config" object which can be used to configure random movement for mobile nodes. You can go through the following steps to set random mobility.

- Create a new scenario called "case5" by duplicating scenario "case1".

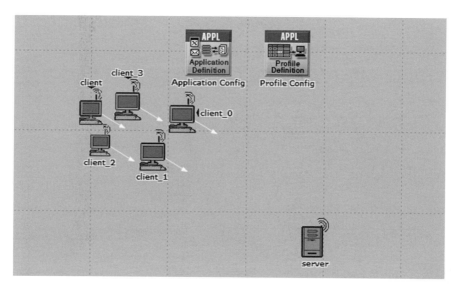

Figure 9.13   Network model

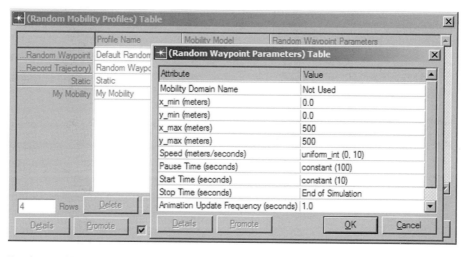

Figure 9.14   Random mobility profiles

- Find the "Mobility Config" object in Object Palette. Place a "Mobility Config" object on the network.
- Right click the "Mobility Config" object, choose "Edit Attributes (Advanced)" to show the "Attributes" dialog. In "Random Mobility Profiles", add a new profile called "My Profile". We can configure the parameters in "My Profile" to produce the expected randomness. In the following example, default parameters are used. The randomness parameter table of "My Profile" is shown in Figure 9.14.
- Select all nodes on the network, choose the menu "Topology" – "Random Mobility" – "Set Mobility Profile...", select "My Profile". All the selected nodes will be configured

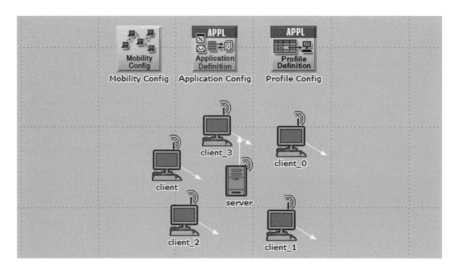

Figure 9.15   Network model

with random mobility based on the randomness set in "My Profile". Figure 9.15 shows the network after configuration.

Now you can choose the statistics in which we are interested and run the simulation. During simulation, these nodes will move randomly based on the randomness profile "My Mobility".

9.4.3   Movement via programming interfaces

In OPNET, node position can be dynamically changed via programming interfaces as well. This method is the most flexible, but one needs to create a custom process model or modify a current process model, because some extra code for dynamically changing positions is required. In OPNET, the position of a communication node consists of three components: "x position", "y position", and "altitude" attributes. We can use the "op_ima_obj_attr_set_dbl()" function to dynamically change the attributes related to node position during simulation. In geo_topo wrapper APIs, there are some useful functions that can be used to facilitate node movement. Consult Chapter 6 for more information.

9.5   Automatic network deployment

In previous sections, you manually deployed network nodes and topologies. Similarly to the automatic deployment tool for wired networks, OPNET Modeler also provides an automatic deployment tool for wireless networks. This deployment tool allows you to deploy a wireless network by specifying: the wireless technology (WLAN, WiMAX, etc.), the network topology (Hexagon, Square), a model of wireless mobile node, and

Figure 9.16 Wireless deployment wizard

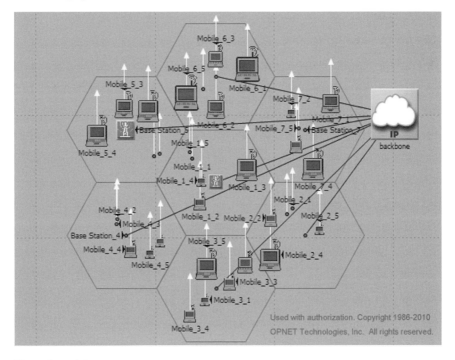

Figure 9.17 Network model

mobility of nodes. In Project Editor, from the "Topology" menu, choose "Deploy Wireless Network..." to show the wireless network deployment wizard as shown in Figure 9.16.

Figure 9.17 shows a hexagon cellular wireless network deployed via the wireless network deployment wizard.

The automatic deployment tool does not deploy network in a very detailed way, but, it does allow fast deployment of large-baseline networks with some common features. More custom configurations can also be made on the automatically deployed baseline networks.

10 Modifying standard models

In this chapter, we will demonstrate how to modify standard models. To follow this chapter, you should understand the modeling concepts and know how to create custom models in OPNET Modeler, as demonstrated in Chapter 5.

10.1 Introduction

In practice, one often needs to design a node or a protocol that functions mostly like a standard node or protocol, but with some special features that a standard node or protocol does not provide. To model this scenario in OPNET Modeler, you can modify the standard model to suit these needs instead of creating the whole model from scratch. All OPNET standard models can be modified, such as node model, process model, link model, packet format model, etc. To modify a model, you need to analyze what you want to modify: simply extend the model's functionality or change the model significantly. You can modify a standard node model by adding a custom module to interact with existing modules in that node. You can modify a standard process model by adding your own code. However, the modification should not influence the correct execution of other code.

10.2 Case study

In this case, you can design a node that functions like a standard PPP (Point to Point Protocol) workstation, which is capable of modifying the IP datagram header's destination address to allow the IP datagram to be routed to a different server from the initially scheduled server. To make such a node model, you need to first open the standard node model "ppp_wkstn_adv" on which the new node model is based. In Node Editor, from "File" menu, choose "Save As..." to save "ppp_wkstn_adv" to "ppp_wkstn_adv_modified". Now you can safely modify the "ppp_wkstn_adv_modified" node model without worrying about changing the original "ppp_wkstn_adv" node model. In "ppp_wkstn_adv_modified" model, add a new module called "extra_layer" and connect it by packet streams to "ip" module, "ip_rx_0_0" module and "ip_tx_0_0" module, as shown in Figure 10.1.

There are four packet streams, as shown in Figure 10.1. These packet streams' "src stream" and "dest stream" attributes should have the following values, so that the code

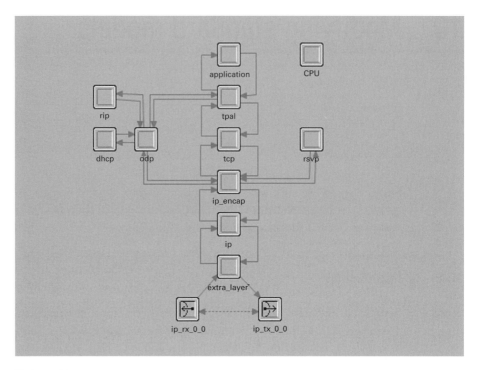

Figure 10.1   Node model

in the corresponding process model can match the correct packet stream input and output ports:

- Packet stream from "extra_layer" module to "ip" module: src stream [0] and dest stream [1]
- Packet stream from "ip" module to "extra_layer" module: src stream [1] and dest stream [0]
- Packet stream from "ip_rx_0_0" module to "extra_layer" module: src stream [0] and dest stream [1]
- Packet stream from "extra_layer" module to "ip_tx_0_0" module: src stream [1] and dest stream [0].

Note: These stream indices can be adjusted by connecting these modules in different orders.

Add a custom "toggle" type attribute called "modify_dest" to the "extra_layer" module and make this attribute "promoted" in order to expose it to the node model, as shown in Figure 10.2.

Next, you can create a process model for the "extra_layer" module. The state transition diagram of this process model is shown in Figure 10.3.

If a packet comes from an upper layer module, in this case the "ip" module, then the state transitions to "upper_in" state. If a packet comes from a lower layer module, in this

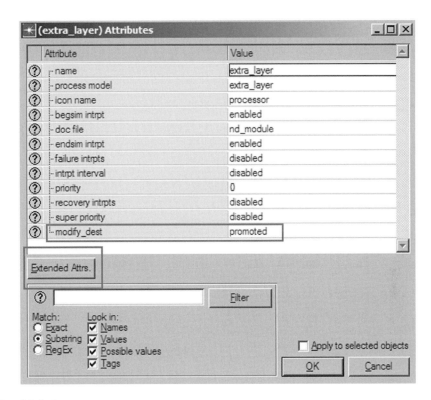

Figure 10.2   Attributes

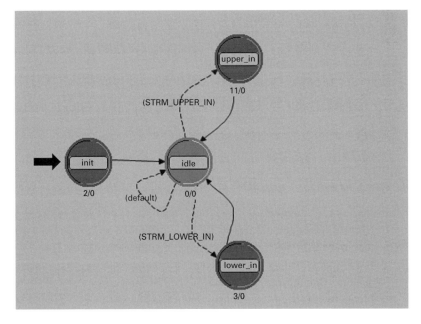

Figure 10.3   Process model

Type	Name	
unsigned int	server_addr	
Objid	node_id	

Figure 10.4   SV block

```
Packet *pk = OPC_NIL;
int modify_dest;
IpT_Dgram_Fields *fields;
unsigned int dest_addr_int;
char ip_str[32];
```

Figure 10.5   TV block

```
#include "ip_dgram_sup.h"
#include "ip_addr_v4.h"
#include <prg_ip_address.h>

#define UPPER_IN_STRM_INDEX 0
#define UPPER_OUT_STRM_INDEX 0
#define LOWER_IN_STRM_INDEX 1
#define LOWER_OUT_STRM_INDEX 1

#define STRM_UPPER_IN ( \
    (op_intrpt_type() == OPC_INTRPT_STRM) && \
    (op_intrpt_strm() == UPPER_IN_STRM_INDEX))
#define STRM_LOWER_IN ( \
    (op_intrpt_type() == OPC_INTRPT_STRM) && \
    (op_intrpt_strm() == LOWER_IN_STRM_INDEX))
```

Figure 10.6   HB block

case "ip_rx_0_0" module, then the state transitions to "lower_in" state. In SV block, add declarations of state variables as shown in Figure 10.4.

In TV block, declarations of temporary variables are added as shown in Figure 10.5.

In HB block, the required header files and define macro conditions used in the state transition diagram are included, as shown in Figure 10.6.

To modify standard models, you should be familiar with the data structures and functions defined in header files in the path "C:\[OPNET installation path]\std\include". This folder contains the data structures and function definitions of OPNET standard protocols and models. In the HB block, include "ip_dgram_sup.h" and "ip_addr_v4.h", which are located in this path. The reason to include the "ip_dgram_sup.h" header file is that you can use both "IpT_Dgram_Fields" data structure and the "ip_dgram_fields_access()" function, both of which are defined in "ip_dgram_sup.h" header file. "IpT_Dgram_Fields" data structure contains the fields of the IP datagram. The "ip_dgram_fields_access()" function allows you to access the "IpT_Dgram_Fields" object in an IP datagram. So you can use this function to read/write the dest_addr member of "IpT_Dgram_Fields" to modify the IP datagram's destination address. It is noted that "src_addr" and "dest_addr" members in "IpT_Dgram_Fields" are "InetT_Address" type, which is defined in the "ip_addr_v4.h" header file, so you should include "ip_addr_v4.h" in the HB block as well. "IpT_Address" is also defined in the "ip_addr_v4.h" header file as unsigned int, so you can manipulate this IP address by using the OPNET IP Address API package, which allows you to convert an IP address between unsigned int and string.

Q10.1   What are the differences between OPNET API functions and the functions defined in standard models' include header files?

OPNET API functions are the basic simulation interfaces that OPNET provides to allow you to model the fundamental simulation elements such as packet, queue, radio, process, interrupt, etc. The functions defined in standard models' include header files are functions that implement some functionalities by utilizing OPNET API functions and provide higher-level interfaces.

HB block also defines four packet stream indices, which correspond to the packet stream src and dest ports for the "ppp_wkstn_adv_modified" node.

In "init" state, add initialization code to this process model, as shown in Figure 10.7.

In "upper_in" state, add the code as in Figure 10.8.

In "upper_in" state, we first check whether the "modify_dest" toggle attribute is enabled in node object. If it is enabled, then the destination IP address of this IP datagram will always be set to the first server IP address. The original destination IP address will be printed out in the simulation console. Finally, this IP datagram will be sent to the lower layer module, in this case, the "ip_tx_0_0" module.

Q10.2   What are the differences between "ip_dgram_fields_access()" and "ip_dgram_fields_get()"?

Both functions can be used to retrieve the IP datagram fields structure from a packet. However, "ip_dgram_fields_access()" only retrieves a pointer to the fields structure, while "ip_dgram_fields_get()" not only retrieves a pointer to the fields structure but also strips off the fields from the packet.

In "lower_in" state, add code to deliver the packet received from "ip_rx_0_0" to the upper layer module, i.e., "ip" module, shown in Figure 10.9.

```
server_addr = 0;
node_id = op_topo_parent(op_id_self());
```

Figure 10.7   "init" state

```
pk = op_pk_get(UPPER_IN_STRM_INDEX);
op_ima_obj_attr_get_toggle(node_id, \
    "extra_layer.modify_dest", \
    &modify_dest);
if(modify_dest == OPC_TRUE)
{
  fields = ip_dgram_fields_access(pk);
  dest_addr_int = fields->dest_addr.address.ipv4_addr;
  // If no previous IP address stored, then store this
  // address as the first IP address
  if(server_addr == 0)
    server_addr = dest_addr_int;
  else
  {
    // Set the IP packet's destination address to
    // the first recorded IP address
    if(fields->dest_addr.address.ipv4_addr != \
        server_addr)
      fields->dest_addr.address.ipv4_addr = server_addr;
  }
  // Print the IP address
  prg_ip_address_value_to_string(dest_addr_int, ip_str);
  op_sim_message( \
      "IP␣destination␣address␣modified", ip_str);
}
op_pk_send(pk, LOWER_OUT_STRM_INDEX);
```

Figure 10.8   "upper_in" state

```
pk = op_pk_get(LOWER_IN_STRM_INDEX);
op_pk_send(pk, UPPER_OUT_STRM_INDEX);
```

Figure 10.9   "lower_in" state

In Process Editor, from the "Interfaces" menu, choose "Process Interface". Set "begsim intrpt" and "endsim intrpt" attributes as "enabled". Save this process model as "extra_layer" and compile it.

Open the "ppp_wkstn_adv_modified" node model in Node Editor. For the "extra_layer" module, set its "process model" attribute to "extra_layer". Save "ppp_wkstn_adv_modified" node model.

Next, you need to create a network to test this modified node model. Create a new project and scenario with project name "chapter10" and scenario name "case1". Add a new subnet object to the scenario in Project Editor, as in Figure 10.10.

Within the subnet object, add the following objects: an "Application Definition" object, a "Profile Definition" object, a "ppp_wkstn_adv_modified" node object named "client", a "slip4_gtwy_adv" node object named "router", and two "ppp_server_adv" objects named "server_a" and "server_b", respectively. Connect "client", "router", "server_a", and "server_b" by using "PPP_DS0" model link objects. This is shown in Figure 10.11.

In "Application Definition" object, add a new Http application called "My Http", as shown in Figure 10.12.

In "Profile Definition" object, add a new profile called "My Profile" containing the "My Http" application, as shown in Figure 10.13.

For a "client" node object, make its "Application: Supported Profiles" attribute include "My Profile", as shown in Figure 10.14.

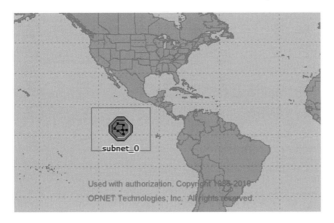

Figure 10.10   Network model

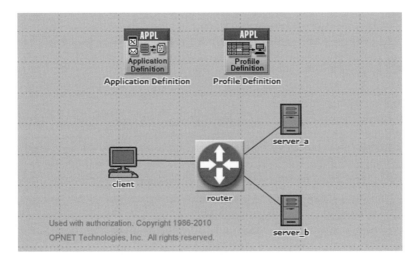

Figure 10.11   Network model

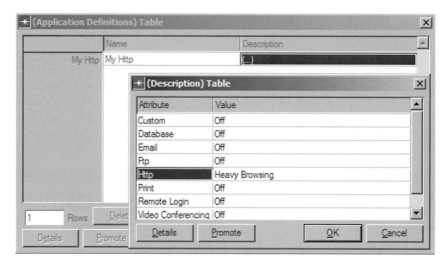

Figure 10.12   Application Definitions

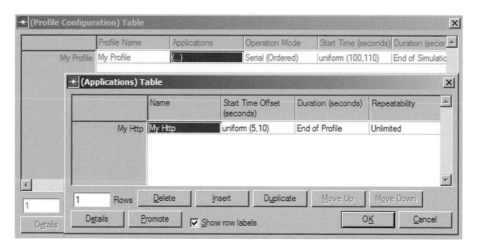

Figure 10.13   Profile Configuration

For both "server_a" and "server_b" node objects, simply make their "Application: Supported Services" support "All", as shown in Figure 10.15.

In this case, you can verify the function of the modified node by using both statistics and animation. For the links of "router" ⟷ "server_a" and "router"⟷"server_b", choose individual DES statistics, as shown in Figure 10.16.

You should choose throughput statistics for these links so that you can see whether there is traffic flowing within these links.

In Project Editor, from the "DES" menu, check "Record Packet Flow 2D Animation For Subnet" to record packet movement in the subnet during simulation. Press "Ctrl+R" and "Alt+R" consecutively to start simulation. After simulation completes, from the

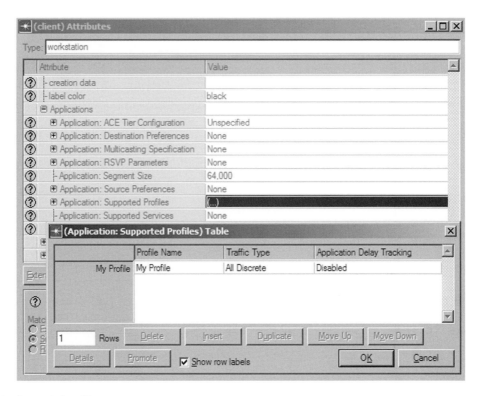

Figure 10.14 Supported profiles

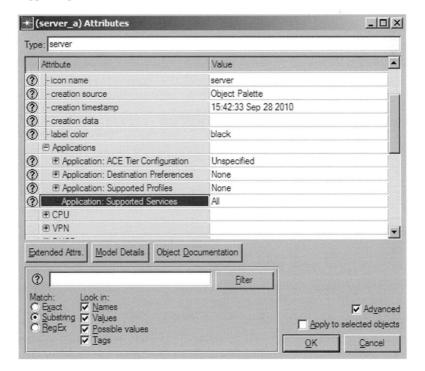

Figure 10.15 Attributes

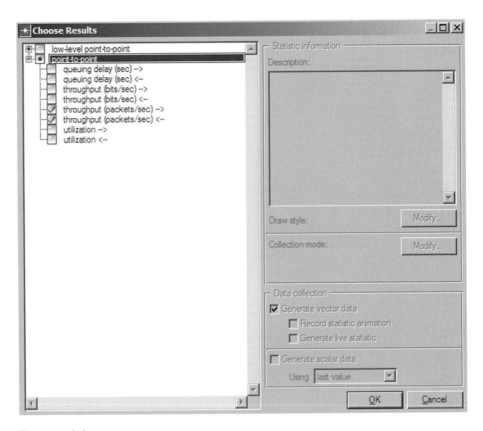

Figure 10.16   Choose statistics

"DES" menu, choose "Play 2D Animation". The packet flow animation shows that the packets are flowing within both "router"←→"server_a" and "router"←→"server_b" links, as shown in Figure 10.17.

We can also double-check it from the statistic results for "router"←→"server_a" and "router"←→"server_b" links. Right click any empty space in Project Editor, in the context menu choose "View Results" to show Results Browser. In the "Object Statistics" list, check the throughput statistics for both links. The results are shown in Figure 10.18. It is seen that the traffic flowing within these two links has similar loads.

Next, you can set "extra_layer.modify_dest" attribute to "enabled" for "client" node object, as shown in Figure 10.19. After setting this attribute to "enabled", the "extra_layer" module within the "client" node object will change the destination IP address, so that all IP datagrams will be routed to one server. Press "Ctrl+R" and "Alt+R" consecutively to run the simulation again. After simulation completes, play the 2D animation for the subnet. It is seen that the packets are flowing within the "router"←→"server_b" link, but not in the "router"←→"server_a" link, as shown in Figure 10.20.

Again, you can inspect the throughput statistics for "router"←→"server_a" and "router"←→"server_b" links, as shown in Figure 10.21.

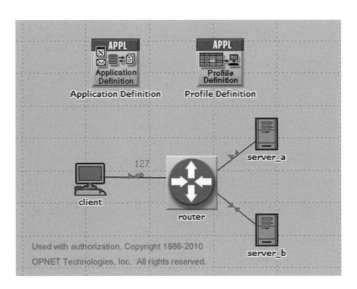

Figure 10.17 Network model

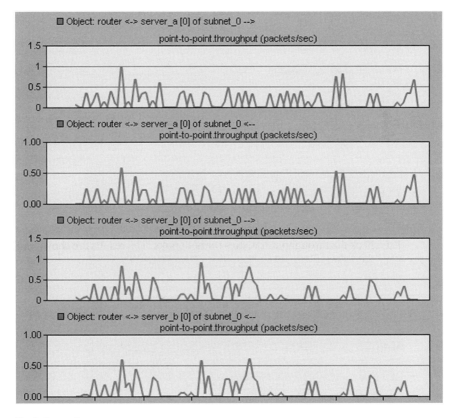

Figure 10.18 Statistic results

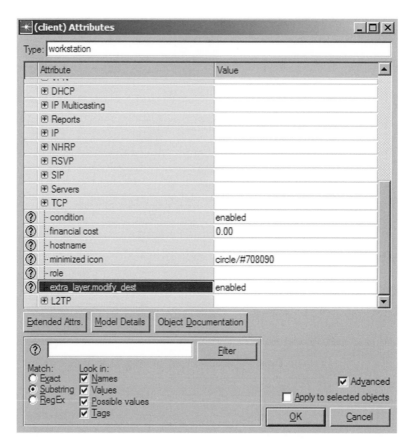

Figure 10.19 Attributes

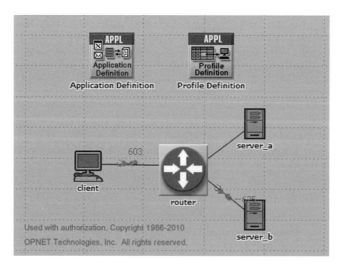

Figure 10.20 Network model

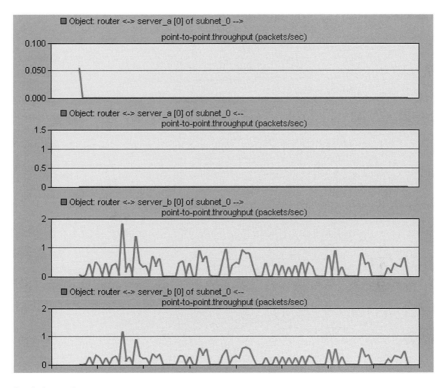

Figure 10.21 Statistic results

It is seen that there is no traffic flowing within the "router"⟷"server_a" link, but the traffic flowing in the "router"⟷"server_b" is nearly doubled. From comparison of the two simulation scenarios in both statistic results and packet flow animations, you can verify that the modified node model functions as expected.

Part IV

OPNET Modeling Facilities

11 Debugging simulation

This chapter describes the debugging facilities that OPNET Modeler provides and shows how to debug OPNET models with different techniques. This chapter assumes that the reader understands general debugging concepts, such as call stack, breakpoint, trace and watch, etc.

11.1 Debugging facilities in OPNET Modeler

OPNET Modeler provides two levels of debugging capability: object-level debugging and source-level debugging. In object-level debugging, *object* refers to an OPNET simulation entity like packet, event, process, etc. The object-level debugging process follows the order in which discrete events are scheduled; therefore, it is normally used when you want to track the simulation on an event-by-event basis and track and inspect the simulation objects associated with these events. Object-level debugging reflects the internal logic of discrete event simulation. In source-level debugging, *source* refers to the simulation source code. The source-level debugging process follows the execution of source code; therefore, it is suitable if you want to track and watch the value of variables and inspect the details of your code. The object-level debugging technique is specific for debugging event-based simulation programs, and the source-level debugging technique is for debugging general software programs. In practice, it is advisable to combine both debugging techniques in order to produce more reliable models.

An object-level debugger is integrated with OPNET Modeler itself. It is called OPNET Simulation Debugger (ODB). For a source-level debugger, most general source code debuggers can be used to debug OPNET programs, such as Microsoft Console Debugger (CDB), Microsoft Visual C++ Debugger (MSVC), and GNU Project Debugger (GDB). CDB and MSVC are source-level debuggers on Windows platforms and GDB is generally used on Linux platforms. Among them, CDB and GDB can be used in conjunction with OPNET debugger, i.e., source code debugging operations can be performed with CDB and GDB within the OPNET debugging window. For MSVC, source code debugging operations are performed within either MSVC's own Integrated Development Environment (IDE) or command line instead of the OPNET debugging window (see www.microsoft.com for Microsoft Visual C++ and Microsoft Console Debugger; www.gnu.org/software/gdb for the GNU Project Debugger).

11.1.1   Prerequisites for debugging

Before debugging simulation, certain preferences need to be set. In the "Edit" menu, choose "Preferences" to show Preferences Editor. The following preferences will be set appropriately:

- Search for "Simulation Kernel Type" preference and set its value to "development". This is to make simulation in development mode so that the simulation program will not be optimized and debugging information, profiling data, and symbols will be reserved.
- For CDB debugging, search for "Show Console Window" preference and set its value to "TRUE".
- For CDB debugging, search for "Path to ??-bit Windows Command-line Debugger" (?? can be either 32 or 64 depending on CPU and OS support on target machine) and set its value to the path of CDB executable file (cdb.exe).
- Search for "Compilation Flags for Development Code" preference. For CDB and MSVC debugging, set its value to "/Zi /Od". For GDB debugging, set its value to "-g". This is to include debugging information in the compiled object file and turn off all optimizations.

It is also assumed that appropriate source-level debuggers such as CDB, GDB, and MSVC have been installed on the target machine.

11.1.2   Preparing simulation scenario

In this section, a simulation scenario will be created in order to demonstrate different debugging techniques. You can first create a new Project with project name "chapter11" and scenario name "case1". In Network Editor of the "case1" scenario, add a node as shown in Figure 11.1.

The node model is "debug_node1". The "debug_node1" node model contains three modules. The "source" and "sink" modules respectively generate and destroy packets and the "processor" module acts as an intermediate module for processing and relaying packets. The node model structure is shown in Figure 11.2.

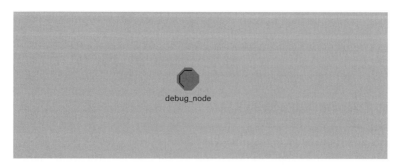

Figure 11.1   Network model

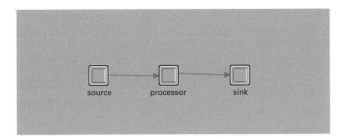

Figure 11.2 Node model

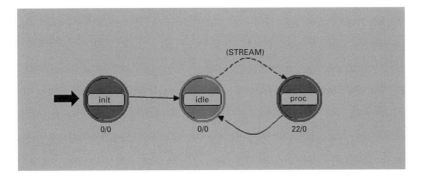

Figure 11.3 Process model

For the "source" module, its "process model" attribute is set to "simple_source" and "Packet Interarrival Time" attribute is set to "exponential (0.001)". For the "sink" module, its "process model" attribute is set to "sink". For the "processor" module, its "process model" attribute is set to "debug_process1", which is a custom process model we create. The logic of "debug_process1" is shown in Figure 11.3.

Figure 11.4 shows the code in the "debug_process1" model. The comments in Figure 11.4 show which block of code should be placed in which state. The "debug_process1" process model needs to be compiled before starting simulation.

In Project Editor, from the "DES" menu, choose "Configure/Run Discrete Event Simulation..." to show the simulation configuration dialog. Expand the "Execution" category from the category list and select "OPNET Debugger" to show debugger settings, as shown in Figure 11.5. Make sure the "Use OPNET Simulation Debugger (ODB)" item is checked so that the OPNET debugging window will show up when running the simulation. This item should be checked for either object-level debugging or source-level debugging.

11.1.3 Debugging with ODB

Since ODB debugging is an object-level debugging, it controls the simulation process in an event-by-event manner, i.e., simulation progresses on completion of events. For

```
/* Header Block (HB) */
#define STREAM (op_intrpt_type () == OPC_INTRPT_STRM)
typedef enum { Destroy, Deliver } pp_status;
static pp_status check_packet (Packet *);

/* Temporary Variables */
int strm_index = 0;
Packet *pkptr = OPC_NIL;
char info_string[32];

/* "proc" state */
op_prg_odb_bkpt ("label_breakpoint1");
strm_index = op_intrpt_strm ();
pkptr = op_pk_get (strm_index);
if(check_packet(pkptr) == Destroy)
{
  if (op_prg_odb_ltrace_active("label_trace1"))
  {
    op_prg_odb_print_major( \
        "=====␣Packet␣to␣Destroy␣=====", OPC_NIL);
    sprintf(info_string, "Size:␣%d", \
        op_pk_total_size_get(pkptr));
    op_prg_odb_print_minor (info_string, OPC_NIL);
    sprintf(info_string, "Creation␣Time:␣%f", \
        op_pk_creation_time_get(pkptr));
    op_prg_odb_print_minor (info_string, OPC_NIL);
  }
  op_pk_destroy(pkptr);
}
else op_pk_send(pkptr, 0);

/* Function Block (FB) */
static pp_status check_packet (Packet *pkptr)
{
  OpT_Packet_Size pkt_size;
  pp_status status;

  FIN(check_packet(pkptr));

  pkt_size = op_pk_total_size_get(pkptr);
  if(pkt_size < 500) status = Destroy;
  else status = Deliver;

  FRET(status);
}
```

Figure 11.4   Code in process model

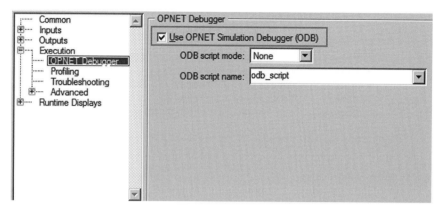

Figure 11.5 Simulation configuration

debugging in ODB, OPNET provides several classes of commands to control the simulation process. To use these commands, the simulation should be in running state. When the simulation starts running, the OPNET debugging window will show up. In the ODB prompt of the OPNET debugging window, type "help" to see all classes of ODB commands available. To view the help for a particular ODB command, you can type "help" followed by the desired command you want to know:

ODB> help next

This will print the usage for the ODB "next" command. With these ODB commands, you can pause simulation at a specified breakpoint, trace the functions used in your model, display memory allocation details, and more. For example, to stop simulation at a specified simulation time, you can type the following commands:

ODB> tstop 22.33

ODB> cont

The "tstop" command sets a breakpoint at the simulation time 22.33 seconds. The "cont" command continues running the simulation. Simulation will be running until the specified breakpoint is reached. Once simulation is paused by hitting the breakpoint, the ODB command can be used to obtain the debugging information, such as printing specified packet information and event details, and so on.

Along with ODB commands, OPNET also provides programming interfaces to facilitate ODB debugging. These interfaces allow you to set labeled breakpoints, trace custom functions, write diagnostic code for the process model, and more.

11.1.3.1 Setting labeled breakpoints

A labeled breakpoint refers to one that is configured in the process model via the op_prg_odb_bkpt() API. Since the labeled breakpoint is configured in the process model, you can accurately place the labeled breakpoint anywhere in the process model's code

```
 _____ (ODB: Event) _____
|                                                                    |
|   * Time   :  564.593167197813 sec, [9m 24s . 593ms 167us 197ns 81|
|   * Event  :  execution ID (1448781), schedule ID (#1448782), type |
|   * Source :  execution ID (1448780), top.subnet_0.debug_node.sour |
|   * Data   :  instrm (0), packet ID (554263)                       |
|   > Module :  top.subnet_0.debug_node.processor [Objid=5] (process |
|                                                                    |
| breakpoint |#1| trapped: "stop at label = (label_breakpoint1)"     |
|_____|
```

Figure 11.6   ODB console output

and make the breakpoint conditional. Therefore, a labeled breakpoint complements the normal breakpoint that is set by ODB commands. The usage of labeled breakpoints is shown in Figure 11.4, where a labeled breakpoint named "label_breakpoint1" is configured in "proc" state by using op_prg_odb_bkpt(). During simulation, labeled breakpoints are not enabled unless they are set to be enabled explicitly via particular ODB commands. To set labeled breakpoint to enabled, type the following ODB command:

ODB> lstop label_breakpoint1

ODB> cont

Then simulation execution will stop at "label_breakpoint1", as in Figure 11.4. The console output after executing these commands is shown in Figure 11.6.

The labeled breakpoint's ID is "#1". This labeled breakpoint can also be deleted using the "delstop" command. In contrast to use of the "lstop" command, the "delstop" command requires the breakpoint's ID as its parameter. In this case, the ID of "label_breakpoint1" is 1.

ODB> delstop 1

The IDs of breakpoints can also be found in the ODB debugging window's "ODB Breakpoints" tab.

11.1.3.2   Tracing user-defined functions

During debugging, one often needs to know the structure of a function and variables in that function. To display information about functions used in simulation models, ODB trace commands can be used to print out the traces of these functions. The trace information includes function name, arguments' names and values, and invocations of other functions. In particular, to display information about user-defined functions via ODB trace commands, some macros should be added to these custom functions. These macros are used in the "check_packet()" function as shown in Figure 11.4. Macro "FIN" should be placed right after declarations of function variables. Macro "FRET" or "FOUT" should be placed at the end of the function. "FIN" marks the start of a user-defined

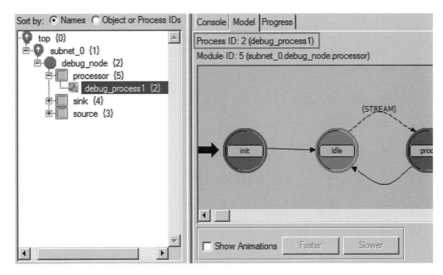

Figure 11.7   OPNET debugging window

function. "FRET" returns a value in a user-defined function. "FOUT" returns void in a user-defined function. After adding these macros to your custom function, you can print the trace of the function by typing the following commands:

$$ODB> \text{protrace} <\text{process id}>$$

$$ODB> \text{cont}$$

<process id> should be replaced by the actual ID of "debug_process1" process model. The ID of a process model can be found in ODB debugging window's "Model" tab, as shown in Figure 11.7.

After executing these commands, the console will print out function traces for the "debug_process1" process model. The trace of the "check_packet()" function will be printed out as well. It is noted that: if "FIN/FOUT/FRET" macros are not used, the function will not be traced.

11.1.3.3   Using diagnostic block

To debug a model, one often expects that some code can be executed to check possible issues and print out relevant information, and the code to be executed when asked. Diagnostic block can be used to serve this purpose. The diagnostic block consists of a sequence of statements which are only executed under the control of ODB commands. Diagnostic block can be used to diagnose model code and print out some useful information if certain conditions are met. In this context, labeled trace is often used alongside ODB commands to trigger the execution of diagnostic block at simulation runtime. Labeled trace is similar to the concept of labeled breakpoint, which can be set to active or inactive by ODB commands. The "op_prg_odb_ltrace_active()" function is used to check whether a labeled trace is set to active by ODB command and to decide whether the diagnostic block should be executed. In Figure 11.4, the labeled trace is "label_trace1". If this labeled trace is activated by ODB command, the diagnostic block will be executed to print out information about packet size and creation time. The labeled trace can be activated and the diagnostic block will be printed out by typing the following ODB commands in OPNET debugging window's ODB command prompt:

ODB> ltrace label_trace1 c

ODB> cont

You can disable the labeled trace using the "susptrace" ODB command or delete the labeled trace using the "deltrace" ODB command. Unlike the "ltrace" command, "susptrace" and "deltrace" accept trace ID as a parameter instead of trace label. Trace ID can be found in the OPNET debugging window's "ODB Traces" tab, as shown in Figure 11.8. In this case, the trace ID for "label_trace1" is 0.

To disable "label_trace1":

ODB> susptrace 0

To delete "label_trace1":

ODB> deltrace 0

With these ODB commands, you can tell simulation to execute code in diagnostic block when it is necessary. Diagnostic block provides the capability of executing certain debugging code for the development simulation kernel and not executing this code for the optimized simulation kernel. Thus, diagnostic block does not impose CPU processing and memory burdens on the final optimized simulation kernel.

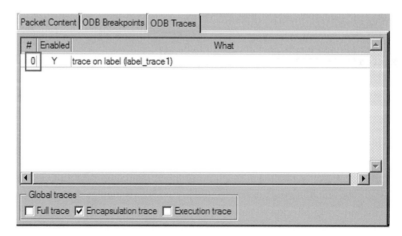

Figure 11.8   OPNET debugging window

11.1.4 Debugging with CDB/GDB

In contrast to ODB debugging, CDB/GDB debugging is source-level debugging. Therefore, CDB/GDB debugging controls simulation process in a statement-by-statement manner, i.e., simulation progresses after executing source code statements. CDB is used for Windows platforms and GDB is generally used for Linux platforms.

First, run the simulation to show the OPNET debugging window. In this window's "Simulation" menu, choose "Attach Windows Debugger (CDB)". Now CDB is attached to simulation and the CDB command prompt is shown in the Console tab, as shown in Figure 11.9.

At this point, the source code is not available. To load source code into the debugging window, you can use the CDB command to trap a breakpoint in the source code. For example, if you want simulation to stop at the first line of "debug_process1" process model, you can set a breakpoint for the "debug_process1" process model by typing the following command in the CDB command prompt:

$$CDB> bp \ debug\_process1$$

$$CDB> g$$

The first command sets a breakpoint for the "debug_process1" symbol and the second continues running the simulation. Then, simulation will stop at the entry point of the "debug_process1" process model and related source code will be loaded into the debugging window, as shown in Figure 11.10.

Now, we can set more breakpoints in the source code and evaluate variable values in the debugging window. To set more breakpoints in the loaded source code, you can set the cursor at a specified line and simply right click the mouse and choose the "Set Breakpoint" menu item, as shown in Figure 11.11.

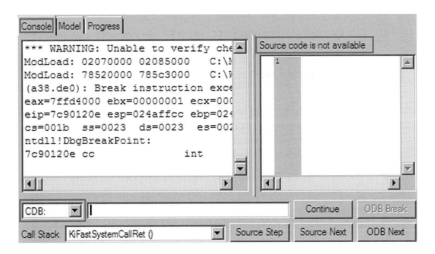

Figure 11.9   OPNET debugging window

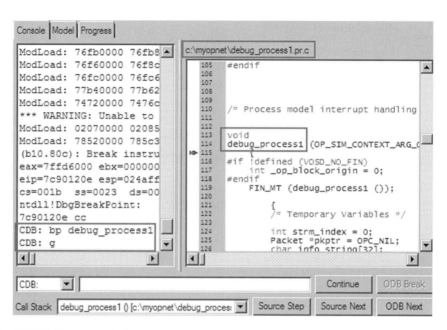

Figure 11.10 OPNET debugging window

Figure 11.11 OPNET debugging window

Similarly, we can disable or delete a breakpoint at a specified line by choosing the context menu "Disable Breakpoint" or "Delete Breakpoint" menu item. Once a breakpoint is trapped, we can evaluate the values of local and state variables in the OPNET debugging window's "Local and State Variables" tab, as shown in Figure 11.12.

Alternatively, you can use CDB commands to control the debugging process as well. For more information on CDB commands, refer to CDB help documentation.

The debugging process for GDB is similar to that for CDB. We should first attach GDB to the simulation by choosing "Attach GDB" in the OPNET debugging window's

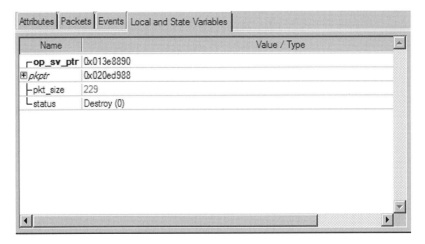

Figure 11.12   OPNET debugging window

"Simulation" menu. Then, you can use CDB commands to trap a breakpoint in order to load source code into the debugging window. In GDB prompt, type the following commands to set breakpoint at the specified source file's specified function:

GDB> break debug_process1.pr.c:debug_process1

GDB> continue

The first command sets a breakpoint at the entry point of the "debug_process1" function in the debug_process1.pr.c source file. Then, simulation will stop at the entry line of the "debug_process1" function and related source code will be loaded into the debugging window. Now you can control the debugging process either via GDB commands, or from the graphic interface by setting breakpoints at specified lines in source code in the same way as in CDB debugging. For more information on GDB commands, refer to GDB help documentation.

It is noted that in the OPNET debugging window you can switch between ODB debugging and CDB/GDB debugging by selecting the corresponding command prompt in the console. Hence, you can easily debug your models at object level and source level at the same time.

11.1.5   Debugging with Microsoft Visual C++ Debugger

Microsoft Visual C++ Debugger (MSVC) is another source-level debugger on Windows. The main difference between MSVC and CDB is that MSVC provides a comprehensive graphic user interface and IntelliSense capability for autodisplay and auto-completion. To debug source code in MSVC, follow these steps:

- Run the simulation to show the OPNET debugging window.
- Start MSVC.

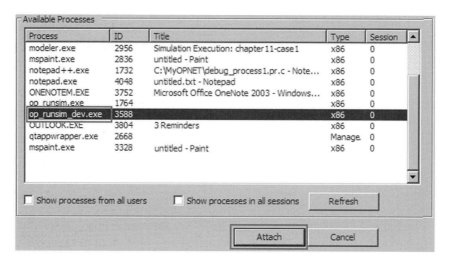

Figure 11.13   Process window

- From MSVC's "Tools" menu, choose "Attach to Process..." to show process window. Here, a process refers to a program loaded into memory by the operating system.
- From the process window, select a process called "op_runsim_dev.exe" and press the "Attach" button to attach Visual C++ Debugger to this selected process, as shown in Figure 11.13.

Q11.1   What is the "op_runsim_dev.exe" process?

"op_runsim_dev.exe" is the finally built simulation kernel process for development. If you set "Simulation Kernel" to "Development", this process will be loaded. The development simulation kernel process provides debugging information. In contrast to the "Development" simulation kernel, there is the "Optimized" simulation kernel, which corresponds to the "op_runsim_opt.exe" process. The optimized simulation kernel process does not provide particular debugging information, but it runs faster than the development simulation kernel process. Simulation kernel type can be set in the "Simulation Configuration" dialog or "OPNET Preferences" dialog.

- In MSVC, you can open all process models' containing source files which you want to debug, then set breakpoints within these source files. For example, to debug source code in the "debug_process1" process model, in MSVC open the containing source file "debug_process1.pr.c", which in this example is located in the "C:\MyOPNET" path. To debug the "simple_source" process model, in MSVC open the "simple_source.pr.c" file from "C:\Program Files\OPNET\[version]\models\std\traf_gen". Now we can set breakpoints in the "debug_process1.pr.c" and "simple_source.pr.c" files.
- In the OPNET debugging window, type the following ODB command to continue simulation:

ODB> cont

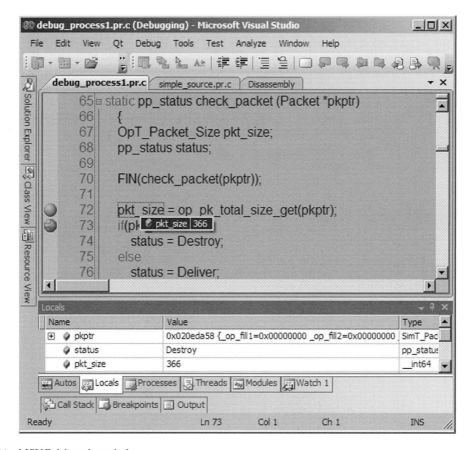

Figure 11.14   MSVC debugging window

Then, simulation will stop at the breakpoints set in the "debug_process1.pr.c" and "simple_source.pr.c" files in MSVC. This is shown in Figure 11.14.

From now on, you can debug these source files in MSVC like normal C/C++ programs by fully utilizing MSVC's powerful debugging user interface, such as by setting breakpoints or conditional breakpoints, evaluating and watching variables, checking call stack, and a lot more. For more information on MSVC, refer to MSVC help documentation.

11.1.6   Debugging with animation

OPNET Modeler provides an animation facility that allows you to visually inspect the packet flows between nodes in a subnet and/or between modules within a node. By utilizing animation, we can easily find communication problems between modules and nodes. To record packet flow animation for the subnet, in Project Editor, from the "DES" menu choose "Record Packet Flow 2D Animation For Subnet". Examples for recording packet flow animation for a subnet can be found in Chapter 5. In this section, we will show how to record flow animation for a particular node. In case1 scenario, select the "debug_node" node, right click this node and select "Choose individual DES statistics" from the context menu. In the dialog, choose "Results Node Animation", as shown in Figure 11.15.

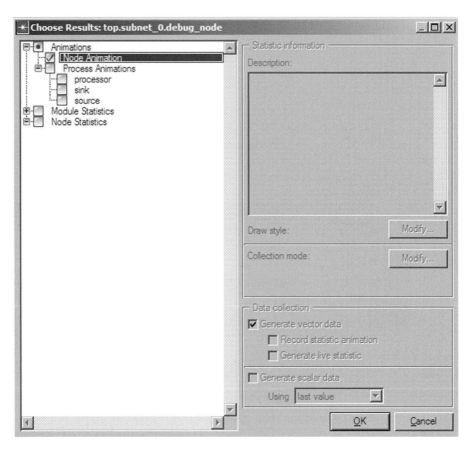

Figure 11.15   "Choose Results" dialog

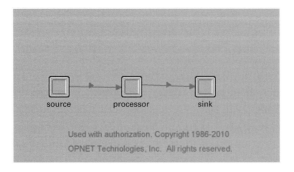

Figure 11.16   Flow animation within node model

Now you can run this simulation scenario. After simulation completes, in Project Editor, from the "DES" menu choose "Play 2D Animation" to play animation for node and process in Animation Browser. Figure 11.16 shows the packet flow animation between modules within the "debug_node" node. If there are communication problems between "source", "processor", and "sink" modules, then there will be no packet flow animation between these modules. Therefore, this animation can help identify communication issues between these modules.

12 OPNET programming in C++

This chapter briefly describes the differences between Proto-C, C, and C++ languages and shows how to configure OPNET to program models in C++. A case study is also provided to help the reader grasp the important points on programming OPNET models in C++. If you are intending to program only in C style, this chapter can be ignored. If you are familiar with C++ and are interested in programming models in object-oriented C++ style, you may read this chapter. The chapter requires basic knowledge of generic C++ programming.

12.1 Proto-C, C, and C++: language and library differences

OPNET provides a programming language called Proto-C to allow users to model various systems. Proto-C preserves generality by incorporating all the capabilities of the C/C++ programming language, i.e., a user can program in Proto-C in a similar way to in C/C++, and in Proto-C you can use the libraries that are accessible to C/C++ as well. On the other hand, Proto-C provides a set of its own APIs to model communication networks. Proto-C supports modeling with the state transition diagram (STD) method, which makes it possible to accurately describe most systems. Figure 12.1 shows that Proto-C allows you to use both C/C++ and Proto-C libraries and write models in both C and C++ styles.

In Figure 12.1, Proto-C APIs, C functions, and C++ methods are mixed together.

Figure 12.2 shows the state transition diagram that models the simplest on/off system. System state transits to on or off depending on the trigger conditions. The Proto-C code can be embedded in each state. If a state is triggered, the Proto-C code within that state may be executed in response. By combining Proto-C and state transition diagrams, complex systems can be modeled.

12.2 Memory management differences between Proto-C APIs and C/C++ standard library functions

Memory can be allocated in either a static way or a dynamical way. In most situations, you declare and allocate memory for a variable in a static way, i.e., at compilation time. However, in some situations, the amount of memory to allocate is known only when

```
strm_index = op_intrpt_strm ();
pktptr = op_pk_get (strm_index);
switch(strm_index)
{
  case 0:
  {
    // Proto-C
    op_pk_send(pktptr, 0);
    // C
    sprintf(str, "%d", ++sent_count);
    // C++
    w.write(std::string("sending␣-␣") + str);
  }
  break;

  case 1:
  {
    // Proto-C
    op_pk_destroy(pktptr);
    // C
    sprintf(str, "%d", ++received_count);
    // C++
    w2.write(std::string("receiving␣-␣") + str);
  }
  break;
}
```

Figure 12.1   Code in process model

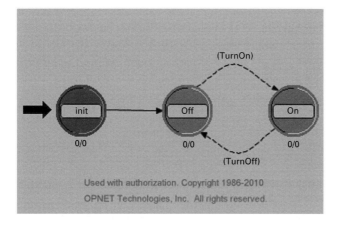

Figure 12.2   State transition diagram

a condition is reached at runtime. For these cases, dynamical memory allocation and a management mechanism are required. In the generic C/C++ programming domain, there are many standard library functions for memory management: C functions such as "malloc()", "realloc()", "memcpy()", and "free()", and C++ new/delete operators. In OPNET programming, these functions and operators can be used to manage memory as well. However, these standard functions and operators provide only generic C/C++ memory management facilities, i.e., no OPNET specific information is provided. Therefore, if there are memory management problems, only generic memory tracking information is provided. This information may be too general to analyze and may not be helpful for resolving the actual problems. Fortunately, Proto-C APIs include a set of memory management functions that manage memory in a dynamical way and provide extra OPNET-specific information and error tracking information. With these memory management functions, if there are memory errors, more specific error tracking information will be provided to help you identify the actual cause.

Proto-C memory functions have three sub-packages for different purposes. These sub-packages are "Memory" sub-package, "Categorized Memory (CMO)" sub-package, and "Pooled Memory (PMO)" sub-package. The memory sub-package includes Proto-C alternatives to C functions like "malloc()", "realloc()", "memcpy()", "free()", etc. You can use APIs in this sub-package to allocate, reallocate, copy, and free memory in a dynamical way, while extra OPNET-specific tracking information is provided for debugging. The Categorized Memory sub-package allows you to define your own memory types and group memory objects into different categories. In this way, memory reports and memory tracking can be more specific. The Pooled Memory sub-package is used to deal with the problem that a large number of fixed-size memory allocation and deallocation operations degrade the performance. This situation happens when you model a repetitive process many times and this process requires dynamical memory allocation and deallocation. If you simply repetitively use memory allocation and deallocation functions to work it out, the performance will degrade to an extent depending on the number of such repetitive operations. The performance is especially important in simulation, because a bottleneck performance degradation may delay the whole simulation process considerably. The Pooled Memory sub-package will help resolve this issue. Repetitive memory allocation and deallocation can be defined via Pooled Memory APIs, and system-specific memory optimization is performed by Pooled Memory APIs to accomplish memory operations.

For details on these Proto-C memory APIs, you can consult the following section in OPNET documentation via documentation browser: "Programmers Reference" – "Discrete Event Simulation" – "Programming Package".

12.3   Proto-C data structures and algorithms packages, C++ standard template libraries (STL) and Boost C++ libraries

To write a simulation model, one often needs to design the protocols for this model. Choosing the right data structure and algorithms can greatly facilitate the design. For

OPNET modeling, there are many options. OPNET Proto-C provides a set of data structures and functions to help you design your algorithms. To design a routing algorithm, you may utilize the DJK package, Geo package, IP Address package, and Graph package. To design a generic algorithm, you may need List package, Mapping package, Hash package, Random package and Vector package. For details on these Proto-C APIs, you can consult the following section in OPNET documentation: "Programmers Reference" – "Data Structures and Algorithms". If you write your model in C++, you can alternatively choose C++ standard template libraries (STL) or more versatile Boost C++ Libraries for designing generic algorithms. In particular, if you want to use Boost C++ libraries, you should add the Boost include path to compilation flags. This include path can be added via Preferences Editor. In Preferences Editor, search the "compilation flags for all code" preference. Add "IC:\Boost\include\boost-?_??" to the end of the preference's value string, where ?_?? should be replaced by the actual version of installed Boost C++ libraries. Now you can include Boost C++ headers in the process model's header block and use Boost libraries in your process models without problems. Similarly, if you want to use some other third-party C/C++ libraries in OPNET, you should set the corresponding compilation flags and/or linking flags to make sure the correct include paths and/or linking paths are set. Users should install Boost libraries first. For instructions on how to install Boost, refer to its official website (www.boost.org).

12.4 Environment configurations for C++ programming in OPNET

With Proto-C, models may be programmed in C style; however, you can also program your models in C++ style. Some configurations must be made before programming in C++ in OPNET. To allow OPNET compiler to compile and link your process model in C++ style, you need to make the following configurations:

- In the process model's header block, add the OPC_COMPILE_CPP flag to the first line. This is to tell OPNET compiler to compile code with C++ awareness. By default, OPNET compiler will compile code only in C.
- In "Edit" menu, choose "Preferences". In the "Preferences" dialog, search the "comp_prog_cpp" tag. You should make sure its value is "comp_g++" for gcc compiler or "comp_msvc" for Visual C++ compiler.
- In the "Preferences dialog", search the "bind_shobj_prog" and "bind_static_prog" tags. For Visual C++ compiler, you should make sure "bind_shobj_prog" tag's value is "bind_so_msvc" and "bind_static_prog" tag's value is "bind_msvc". These two preferences are used to specify which linker programs should be used for creating dynamical and static model libraries.

From now on, you can write your models in C++ style in your process models. All C++ include files should be added to header block. The C++ classes can be declared and defined in the same way as structure data types. Figure 12.3 shows definitions and declarations of both structure and C++ class. In Figure 12.3, "writer" is a normal structure

```
// Define structure and class in header block
OPC_COMPILE_CPP

#define STREAM (op_intrpt_type () == OPC_INTRPT_STRM)

#include <iostream>
#include <string>

struct writer
{
  void write(const std::string &str)
  {
    std::cout << str << std::endl;
  }
};

class writer2
{
  public:
  void write(const std::string &str)
  {
    std::cout << str << std::endl;
  }
};
// Declare structure and class objects in temp block
writer w;
writer2 w2;
```

Figure 12.3   Structure and class

and "writer2" is a C++ class. They are defined in the header block and declared in the temp block in the same way.

The work flows for writing models in C and in C++ are similar. You can just treat C++ class in the same way as structure. However, once you turn on C++ support in OPNET, you can declare temporary variables directly within states, i.e., you do not need to declare temporary variables in "TV" block. Figures 12.4 and 12.5 shows this difference.

Figure 12.4 shows how to declare temporary variables and write code in state in C style. Figure 12.5 shows how to declare variables and write code in C++ style. It is noted that with C++ support turned on, you can declare variables and write code straight away in states. However, for C style, you have to declare variables in TV block and use these variables in states. Therefore, from the point of view of programming flexibility, it is convenient to turn on C++ support.

```
// Declare temporary variables in temp block (TV)
int size = 0;
int i = 0;
Packet *pktptr = OPC_NIL;

// Write process code in state
for(i = 0; i < 10; ++i)
{
  size += i;
  pktptr = op_pk_create(1024);
  op_pk_send(pktptr, 0);
}
```

Figure 12.4 Variable declarations

```
// Declare temporary variables in state
// Write process code in state
int size = 0;
for(int i = 0; i < 10; ++i)
{
  size += i;
  Packet *pktptr = op_pk_create(1024);
  op_pk_send(pktptr, 0);
}
```

Figure 12.5 Variable declarations

12.5 Case study on programming OPNET models in C++

In this case study section, we will use a simple case to help users understand how to write C++ codes and use third-party C++ libraries in OPNET. The C++ libraries used in this case study include those from both STL and Boost.

In this case study, there is a network with two identical nodes connecting each other. These two nodes send unformatted packets to each other. The packet size and interarrival times are not important in this case, i.e., you can set arbitrary values. However, in every packet there is one structure field which carries a string. This string is actually a sequence of many random numbers concatenated by "@" symbols. For example, a sequence may look like: "121@101@32@15@68@100@137". You can also define a set of commands which are represented as numbers. The task is: any node receiving a packet sent by the other node should check whether the sequence carried by this packet contains one or more numbers that are the same as the defined commands; if so, pick them up and print them in console. Otherwise, print "No Command" in console.

Figure 12.6 shows the network domain topology for this case model.

Figure 12.7 shows the node domain modules. There are four modules: transmitter, receiver, traffic generator, and processor. For traffic generator module, the process model is "simple_source" and the packet format is "NONE". For transmitter and receiver modules, set data rate to "unspecified" and packet formats as "all formatted, unformatted". For processor module, we need to create a custom process model called "command_parser", which has three states: "init", "idle", and "process". Figure 12.8 shows the state transition diagram of the "command_parser" process model.

The process starts from "init" state, where initialization codes can be placed. Process control stays in "idle" state until a packet arrives. The packet arrival interrupt is defined as "STREAM" condition. Packet arrival interrupt will cause process transits from "idle" state to "process" state, where the packet will be processed: sequence will be generated and parsed.

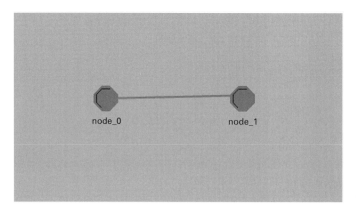

Figure 12.6   Network domain topology

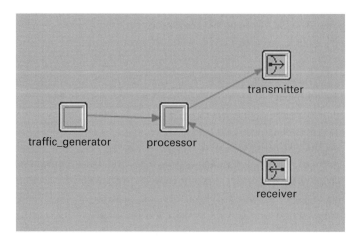

Figure 12.7   Node domain modules

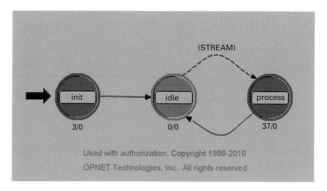

Figure 12.8 State transition diagram for "processor" module

Type	Name
command_parser	parser
sequence_generator	generator

Figure 12.9 State variables

```
OPC_COMPILE_CPP

#include <iostream>
#include <string>
#include <vector>
#include <sstream>
#include <boost/tokenizer.hpp>


#define STREAM (op_intrpt_type () == OPC_INTRPT_STRM)
```

Figure 12.10 Code in header block (HB)

In "command_parser" process model, two C++ classes are defined. One class is "sequence_generator", which is used to generate a sequence containing a number of arbitrary integer numbers. These numbers are concatenated by "@" symbols. Another class is "command_parser", which is able to parse a sequence into a list containing all numbers within that sequence. "command_parser" also checks if this list contains one or more numbers that are equal to the numbers of predefined commands. If this sequence contains one or more command numbers, then these command numbers will be printed out in console; otherwise, "No Command" will be printed out.

Two state variables (SV) are declared, as shown in Figure 12.9.

Figure 12.10 shows the first several lines in the header block (HB). The first line turns on support for C++. The following lines include the required headers for third-party C++ libraries. The last line defines the packet arrival condition that triggers state transition.

```
struct data
{
  std::string sequence;
};

class sequence_generator
{
  Distribution *number_dist;
  std::string sequence;
  int number_count;

  public:
  sequence_generator(int start = 0, int end = 200, \
      int number_count = 20)
  {
    number_dist = op_dist_load ("uniform_int", \
        start, end);
    this->number_count = number_count;
  }
  ~sequence_generator()
  {
    op_dist_unload(number_dist);
  }
  std::string& generate()
  {
    sequence.clear();
    for(int i = 0; i < number_count; ++i)
      sequence += to_string( \
          op_dist_outcome (number_dist)) + "@";

    return sequence;
  }
  template <class T>
  inline std::string to_string(const T& val)
  {
    std::stringstream strstrm;
    strstrm << val;
    return strstrm.str();
  }
};
```

Figure 12.11   Code in header block (HB)

```cpp
class command_parser
{
  std::vector <std::string>commands;

  public:
  typedef enum
  {
    command1 = 100,
    command2,
    command3
  } command_type;
  std::vector <std::string> &parse( \
      const std::string &data, \
      const std::string &delims)
  {
    typedef boost::tokenizer \
        < boost::char_separator<char> > tokenizer_type;
    boost::char_separator<char> sep(delims.c_str());
    tokenizer_type parser(data, sep);
    commands.clear();
    for(tokenizer_type::iterator it = parser.begin(); \
        it != parser.end(); ++it)
    {
      int number = atoi(it->c_str());
      if(number >= command1 && number <= command3)
        commands.push_back(*it);
    }

    return commands;
  }
};
```

Figure 12.12 Code in header block (HB)

Figure 12.11 defines "data" structure and "sequence_generator" class. By default, this class generates 20 random integers between 0 and 200, and concatenates them together by "@" symbols to produce a sequence. This sequence is a string stored in the "data" structure which will be associated with packets from the packet generator.

Figure 12.12 defines "command_parser" class. This class is able to parse a sequence string into a list of integer numbers, and check if this list contains numbers that are equal to numbers of predefined commands. Within this class, three commands are defined. They are numbered 100, 101, and 102. Therefore, if a sequence contains one or more numbers that are equal to 100 or 101 or 102, this sequence contains commands.

```cpp
int strm_index = op_intrpt_strm ();
Packet *pktptr = op_pk_get (strm_index);

myFunc(pktptr);

data *d = OPC_NIL;
switch(strm_index)
{
  case 0:
  {
    d = (struct data *) op_prg_mem_alloc( \
        sizeof (data));
    d->sequence = generator.generate();
    op_pk_fd_set(pktptr, 0, \
        OPC_FIELD_TYPE_STRUCT, d, 0,
        op_prg_mem_copy_create, op_prg_mem_free, \
        sizeof(data));
    op_pk_send(pktptr, 0);
  }
  break;

  case 1:
  {
    op_pk_fd_get(pktptr, 0, &d);
    std::cout << "--------------------------" \
        << std::endl;
    std::cout << "Packet␣Id:␣" << op_pk_id(pktptr) \
        << std::endl;
    std::vector <std::string>& commands = \
        parser.parse(d->sequence, "@");
    if(commands.size() > 0)
    {
      for(std::vector <std::string>::iterator it = \
          commands.begin(); it != commands.end(); ++it)
          std::cout << "Command:␣" << *it << std::endl;
    }
    else std::cout << "No␣Command" << std::endl;
    std::cout << "--------------------------" \
        << std::endl;
    op_pk_destroy(pktptr);
  }
  break;
}
```

Figure 12.13   Code in "process" state

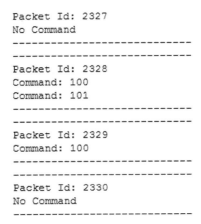

```
Packet Id: 2327
No Command
---------------------------
---------------------------
Packet Id: 2328
Command: 100
Command: 101
---------------------------
---------------------------
Packet Id: 2329
Command: 100
---------------------------
---------------------------
Packet Id: 2330
No Command
---------------------------
```

Figure 12.14   Console output

Figure 12.13 shows the code in "process" state. "process" state will be triggered if a packet arrives. If the packet comes from a stream indexed 0, the packet is sent by the "traffic_generator" module and it should be sent to the transmitter module via "op_pk_send". Before it is sent to transmitter, a sequence is generated and set into a field of this packet. If the packet comes from a stream indexed 1, the packet is sent by the other node. For this case, the sequence will be extracted from the packet and parsed to find out whether this sequence contains command numbers. If it contains one or more command numbers, these numbers will be printed out in console; otherwise, "No Command" text will be printed out. After that, this packet is destroyed.

This model is completed; now you can run the simulation. In order to see the output, you should run your simulation in debugging mode, which allows console output to be displayed. To enable the simulation debugger, in "Configure/Run DES" dialog, in "Execution" – "OPNET Debugger", tick the "Use OPNET Simulation Debugger (ODB)" checkbox. Figure 12.14 shows the console output.

This case study demonstrates how to turn on C++ support, how to include and use third-party libraries, and how to write OPNET models in C++ in general. You can modify this model to allow packets to carry any information of interest and parse them appropriately to achieve some modeling tasks. The information carried by packets can have real-world entities or can be just simulation control information.

13 Traffic in OPNET simulation

This chapter shows how to generate traffic in OPNET Modeler in different ways, including explicit traffic and background traffic. To follow this chapter, it is assumed that a reader knows the basic concepts and operations of OPNET modeling.

13.1 Introduction

In Chapters 8 and 9, we demonstrated how to generate traffic based on application and profile models. However, it is also possible to generate traffic based on the traffic characteristics (packet interarrival times and packet size distributions) rather than applications. The custom models demonstrated in Chapter 5 generate traffic in this way. You can generate traffic this way with standard models as well. Further, you can generate self-similar traffic, background traffic which is based on analytical model, and even hybrid traffic which combines both explicit traffic and background traffic.

From the perspective of simulation methodology, traffic in OPNET modeling can be categorized into two groups: explicit DES traffic and background traffic. Explicit traffic includes traffic based on application model, traffic based on traffic generation parameters, and self-similar traffic based on a raw packet generator (RPG) model. Background traffic includes traffic based on baseline load and traffic based on traffic flow. In the following sections, we will demonstrate how to generate these different types of traffic in OPNET Modeler.

13.2 Explicit traffic

Explicit traffic models traffic in a packet-by-packet basis. It models packet creation, transmission, queuing, and destruction explicitly through a discrete-event simulation process. Therefore, explicit traffic can accurately model the details of protocols. This is the default option for modeling. Explicit traffic can be used to model every detail in discrete-event simulation. However, it takes more time and more resources. In order to reduce the computational burden, background traffic can be used, which will be introduced in the next section. In this section, we demonstrate how to generate explicit traffic in different ways for different uses.

13.2.1　Explicit traffic based on application

Create a new project scenario with project name "chapter 13" and scenario name "case1". Add several models to the Project Editor, as shown in Figure 13.1.

"Application Definition" object's model is the "Application Config" type. "Profile Definition" object's model is of "Profile Config" type. "client" object's model is of "ppp_wkstn_adv" type. "router" object's model is of "slip2_gtwy_adv" type. "server" object's model is of "ppp_server_adv" type. The model of the links between "client" and "router", and "router" and "server" is "PPP_DS1".

For Application Definition, an application called "My Http" is defined, as shown in Figure 13.2.

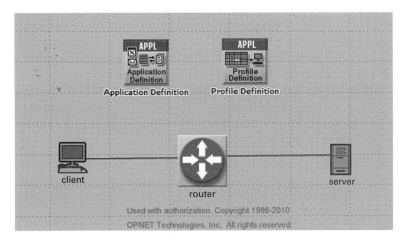

Figure 13.1　Network model

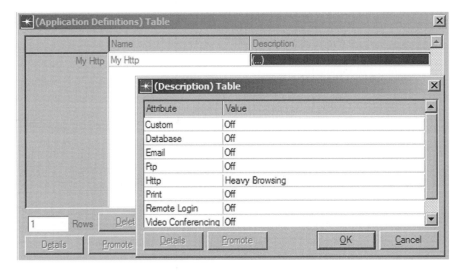

Figure 13.2　Custom node model

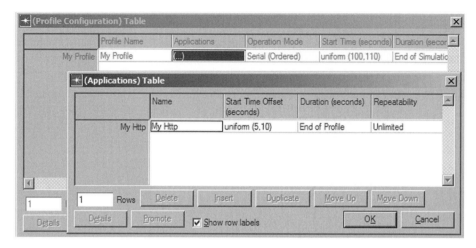

Figure 13.3   Profile Configuration

Figure 13.4   Application Supported Profiles

For Profile Definition, a profile called "My Profile" including "My Http" application is defined, as shown in Figure 13.3.

For "client" node object, "Application: Supported Profiles" is set to "My Profile", as shown in Figure 13.4.

For "server" node object, its "Application: Supported Services" is set to "All", as shown in Figure 13.5.

Now you can run the simulation. The traffic based on application will be generated during the simulation.

13.2.2   Explicit traffic based on traffic generation parameters

In practice, one often wants to generate traffic that has a certain data rate and follows a certain distribution. Traffic based on application cannot guarantee the data rate and traffic distribution. However, OPNET Modeler also provides a set of standard models that are capable of generating traffic which can be parameterized by packet interarrival time and packet size distributions. In OPNET Modeler, the node models that end with

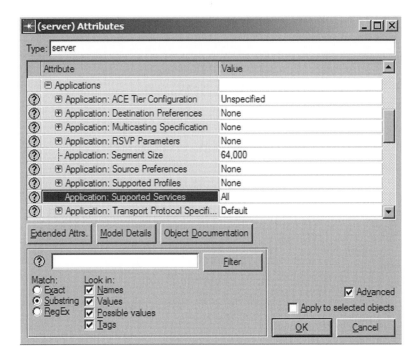

Figure 13.5   Attributes

"_station" and "_uni_src" can be used to generate parameterized traffic. Some of these nodes include "atm_uni_src", "ethernet_station", "ppp_ip_station", and "fddi_station". These nodes can be found in Object Palette. Next, we will demonstrate how to generate parameterized traffic with these node models, based on the case1 scenario.

In Project Editor, from the "Scenarios" menu, choose "Duplicate Scenario..." to save the duplicated scenario as "case2". Change both "client" and "server" objects' "model" attributes to "ppp_ip_station". Now you can edit the "Traffic Generation Parameters" attribute of both "client" and "server" nodes to generate parameterized traffic. This attribute can be found in "IP" – "Traffic Generation Parameters". For both "client" and "server" nodes, in the "Traffic Generation Parameters" table, add a new traffic row and set the packet interarrival time and packet size distributions, as shown in Figure 13.6. You can adjust these traffic parameters to control the traffic data rate.

Now you can run the simulation. The parameterized traffic will be generated during the simulation.

13.2.3   Explicit self-similar traffic based on raw packet generator (RPG) model

If you want to model more bursty traffic like self-similar traffic, you can use a raw packet generator model which models the Fractal Point Processes (FPPs) such as Sup-FRP, PowON-PowOFF, PowON-ExpOFF, ExpON-PowOFF, etc. RPG models are implemented over IP and MAC layers, including "ppp_rpg_wkstn" and "ethernet_rpg_wkstn"

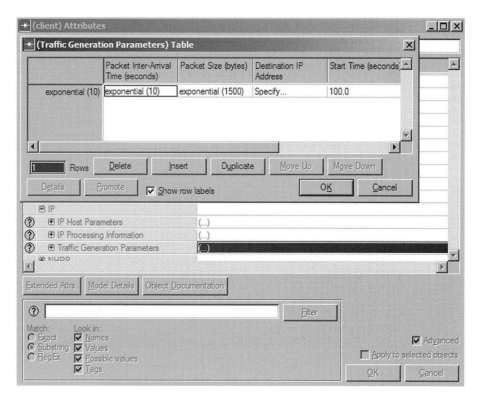

Figure 13.6   Traffic Generation Parameters

(Ryu and Laven, 1998, Ryu, 2000). Therefore, you can generate IP and Ethernet traffic with self-similarity.

To generate such self-similar traffic via RPG models, you need a scenario with RPG nodes. In Project Editor, from "Scenarios" menu, choose "Duplicate Scenario..." to save the duplicated scenario as "case3". Change both "client" and "server" objects' "model" attributes to "ppp_rpg_wkstn_adv". Now you can edit the "RPG Traffic Generation Parameters" attribute of both "client" and "server" nodes to generate parameterized RPG traffic, as shown in Figure 13.7.

In Figure 13.7, you can choose the "Sup-FRP" FPP arrival process, but you can also choose other arrival processes. It is possible to edit the parameters of the FPP arrival process. To do that, click the arrival process to show "Arrival Process" table, where you can change the FPP process parameters such as "Hurst" and "Peak-to-Mean Ratio", as shown in Figure 13.8.

In the "RPG Traffic Generation Parameters" table, click the "Destination Information" item; in the "Destination Information" table, click the "Destination Name" item, for the "Node Name" attribute; choose "subnet_0.server" if current node is "client", otherwise, choose "subnet_0.client", as shown in Figure 13.9.

Now you can run the simulation. The parameterized RPG traffic with self-similarity will be generated during the simulation.

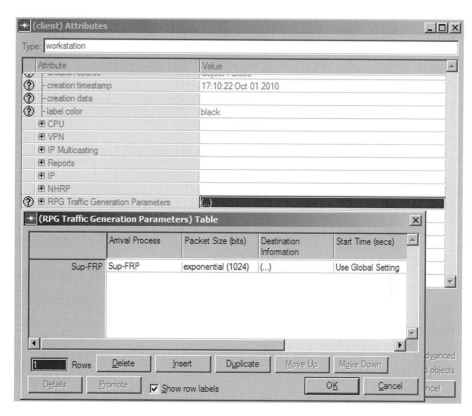

Figure 13.7   RPG Traffic Generation Parameters

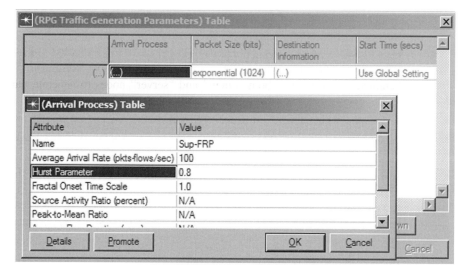

Figure 13.8   Arrival Process

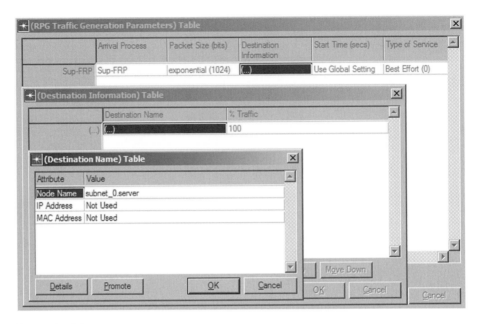

Figure 13.9   Destination Name

13.3   Background traffic and hybrid simulation

Traffic is generally modeled explicitly in discrete-event simulation because explicit traffic modeling allows detailed control of the modeled protocol and provides more accurate results. However, modeling every detail is a time-consuming and resource-consuming process. Modeling explicit traffic in a large network can take a very long time and require more computing resources. In this situation, background traffic can be helpful. Background traffic is modeled via analytical means rather than discrete-event simulation; therefore, it takes much less time and consumes a very small amount of computing resources. You can combine explicit traffic and background traffic in your simulation model. For this case, the performance of explicit traffic is actually affected by the additional delays analytically calculated based on the background traffic load. This is also called hybrid simulation. It is noted that the background traffic cannot be used for all simulation scenarios, since it is based on analytical models; thus only statistics related to the introduced delays may be analytically modeled and other statistics that cannot be derived from the delays will not be affected by background traffic. The implementation of background traffic is limited to IP traffic, statistics of protocols at higher layers than IP, such as TCP, UDP, and applications, are not affected by background traffic modeling. However, OPNET is improving its capability: always check OPNET documentation for updated information on background traffic. In this section, we will demonstrate how to generate background traffic in baseline load and in traffic flow, and how to combine both explicit traffic and background traffic in the simulation as a compromise between accuracy and efficiency. Simulation by combining both explicit traffic and background traffic is also called hybrid simulation.

13.3.1 Background traffic based on baseline load

A baseline load is a static throughput. A baseline load can be configured only for one object from its attribute. The object can be a link model or a node model.

In Project Editor, from the "Scenarios" menu, choose "Duplicate Scenario..." to save the duplicated scenario as "case4". For "client" node and "server" node, choose the "RPG – Delay (secs)" statistic, as shown in Figure 13.10.

The simulation for the case4 scenario can now be run. After simulation completes, statistic results for "client" node or "server" node can be viewed. The RPG delay statistic result is shown in Figure 13.11.

In Project Editor, from the "Scenarios" menu, choose "Duplicate Scenario..." to save the duplicated scenario as "case5". Next, we will add baseline loads for links between "client" and "server" nodes in case5. First, select the link between "client" and "router",

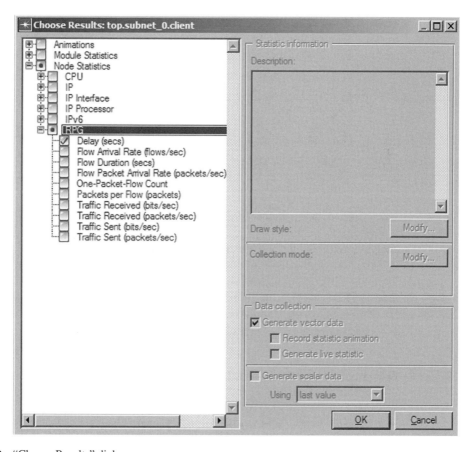

Figure 13.10 "Choose Results" dialog

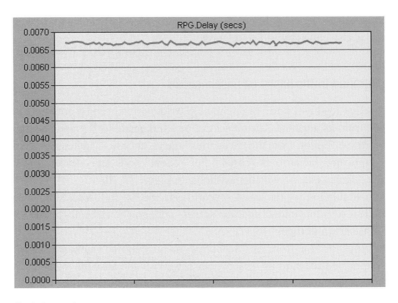

Figure 13.11   Statistic results

edit the "Traffic Information" attribute in the "Traffic Information" table, edit baseline load for "subnet_0.client→subnet_0.router" direction to show "A→B" table, as shown in Figure 13.12. Also edit baseline load for the "subnet_0.router→subnet_0.client" direction in the same way.

In "A→B" table, click the "Traffic Load (bps)" attribute to add a new baseline load profile. In the traffic load profile, add background traffic loads for three intervals along the simulation time axis, as shown in Figure 13.13.

Set the baseline load traffic for the link between "router" to "server" in the same way. Run the simulation. After the simulation completes, view statistic results for "client" node or "server" node. The RPG delay statistic result is shown in Figure 13.14.

It is noted that the RPG delay statistic resonates with the background traffic load set in link objects. Compared with Figure 13.11, the RPG delay with background traffic is longer than without background traffic. In this simulation scenario, the traffic generated includes both explicit traffic from RPG nodes and background traffic within links. Therefore, it is a hybrid simulation scenario. The background traffic set in the links ranges from 100 Kbps to 500 Kbps. If we model this volume of traffic explicitly in RPG nodes rather than in links via baseline load, then it takes a lot longer to complete the simulation and more computing resources are required for this simulation.

13.3.2   Background traffic based on traffic flow

In case4, you can change baseline load to model the background traffic within the links between "client" and "server" nodes. However, if there are many links between "client" and "server" nodes, then you can change the baseline loads for all these links en route. Every time you model different background traffic, you have to change the baseline loads

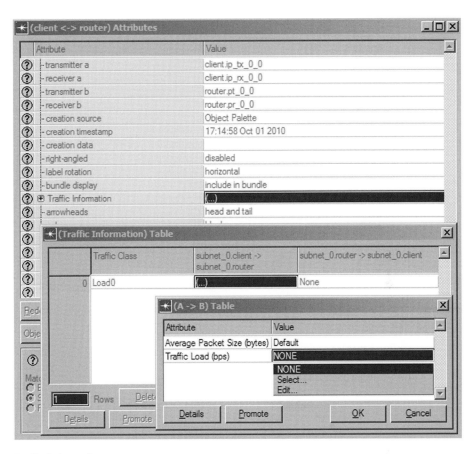

Figure 13.12   Traffic Information

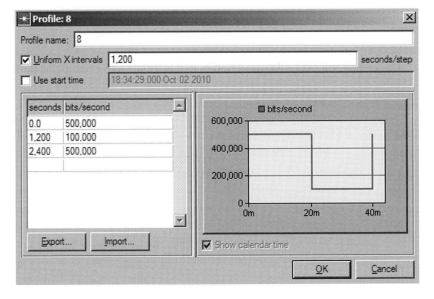

Figure 13.13   Baseline load profile

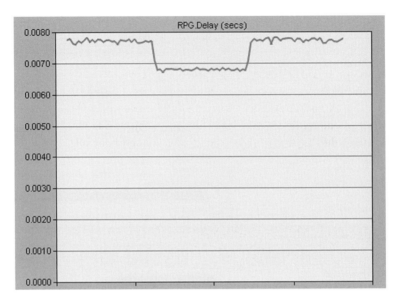

Figure 13.14   Statistic results

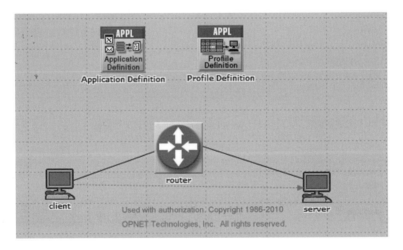

Figure 13.15   Network model

for all links as well. To reduce the work involved with configuring baseline loads for all links en route, OPNET Modeler provides another model to help configure background traffic in a unified way: Traffic Flow model. With this model, you just need to configure the background traffic load once in the Traffic Flow model object, then all objects along the flow path will have that background traffic load automatically.

Open the case4 scenario, from "Scenarios" menu, choose "Duplicate Scenario..." to save the duplicated scenario as "case6". It is noted that in case4, the background traffic is not configured for the links between "client" and "server". In case6, you can use the "Traffic Flow" object to configure the background traffic for all objects between "client" and "server". In the Object Palette, find the "ip_traffic_flow" model. Connect "client" and "server" by using the "ip_traffic_flow" object, as shown in Figure 13.15.

"ip_traffic_flow" is a simplex model. You can also connect "server" and "client" to model duplex background traffic between these two nodes. Next, you can configure the background traffic via the "ip_traffic_flow" object. Edit the "Traffic (bits/second)" attribute of the "ip_traffic_flow" object. For this attribute, add a profile as shown in Figure 13.16.

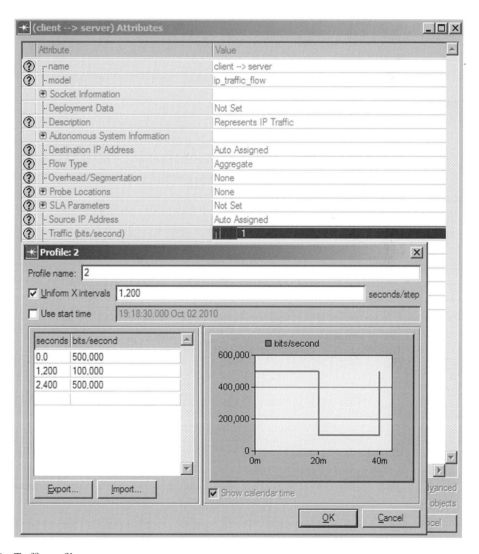

Figure 13.16 Traffic profile

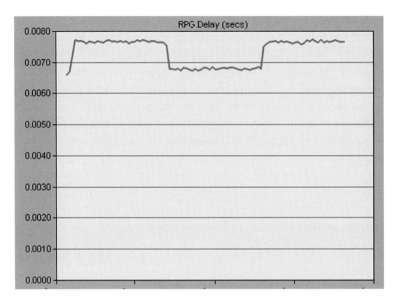

Figure 13.17   Statistic results

It is seen that setting background traffic in the "Traffic Flow" object is the same as in the link object. However, for the case of Traffic Flow, you need to configure it for only one object without worrying about how many objects are between the "client" and the "server".

Now you can run the simulation. Figure 13.17 shows the RPG delay statistic for "client" node. The RPG delay statistic is similar to that in Figure 13.14.

14 External model access (EMA)

This chapter shows the concept of external model access (EMA; a text file representation alternative to OPNET modeler's graphic representation for modeling), the benefits of using EMA in some circumstances, and how simulation models can be constructed via EMA instead of OPNET Modeler's graphic interfaces.

14.1 What EMA is and reasons to use it

Usually, a user can access OPNET models by creating a node model or link model and reading/writing that model from OPNET modeler's graphic user interfaces (GUI) such as Node Editor, Link Editor, etc. This way can be called WYSIWYG – "What You See Is What You Get." Differing from this WYSIWYG method, EMA is a technique provided by OPNET Modeler that allows you to access OPNET models from external programs in a text format; i.e., models can be accessed via code rather than GUI. EMA code can be written in an external C/C++ program. It is different from the C/C++ code written in OPNET process model, which is via the OPNET graphic interface. Therefore, the external C/C++ program with EMA capability can interface with other programs, libraries, and databases just like a general C/C++ program. The EMA C/C++ program is compiled and linked into an executable file. By running that executable file, models and/or objects of models can be read or written or created. To write such an EMA program, you can use any text editor or C/C++ IDE (Integrated Development Environment). The models which can be accessed and created via EMA code include most of the OPNET Modeler model types such as project model, network model, node model, process model, link model, etc. For a full list of model EMA API supports, consult OPNET documentation "Programmers Reference", "Model File Access" section.

OPNET Modeler provides both GUI and EMA methods to allow users to access its models, such as create model, read and write model attributes and so on. However, GUI and EMA have their own advantages and disadvantages. With GUI, the user can create prototype models or read/write models' attributes in an intuitive way without the need for any programming knowledge or EMA-specific knowledge and issues. The GUI method can be applied in most situations. However, in some circumstances, using the GUI method is awkard. For example, in a wireless network there are many wireless nodes and you want to change the power attribute of these nodes' transmitters in two scenarios. In scenario one, all nodes' power attributes should have different power values.

In scenario two, all nodes' power attributes values should follow a particular distribution model. With the GUI method, for scenario one you can change the power attributes of these nodes one by one from the attribute dialog; however, this is tedious if there are many nodes in the network. For scenario two, it is more difficult to achieve the task with the GUI method. Considering a network with hundreds of nodes, modifying attributes one after another takes too much time. The EMA method provides an easier and more scalable solution for these scenarios. With EMA, you can easily loop through objects to change the power attributes of all nodes via programming interfaces. However, in practice, you may combine the benefits of both GUI and EMA methods to access OPNET models.

14.2   EMA case study

You may create a model completely via the EMA method. However, it is easier to create a basic prototype model via the graphic interface and add customized features via EMA. Combining both GUI and EMA methods reduces the burden of writing the same routines every time. In this example, we will create a network model that contains 100 nodes, each of which has a uniformly distributed random altitude ranging from 0 to 1000 meters. This network model may simulate a wireless sensor network (Sohraby et al., 2007) with sensor nodes spread over an uneven area.

First, you can create a node model in Node Editor, within which a "processor"' module is connected to a radio receiver module and radio transmitter module. This node model is saved as "ema_node1". It is shown in Figure 14.1.

Secondly, you can create a new project with project name "chapter14" and scenario name "case1". A subnet object is placed on the network. The subnet contains only one instance of the "ema_node1" node model. Save this project scenario. This is shown in Figure 14.2.

Next, you can generate the EMA code from the current network model. From the Project Editor's "File" menu, choose "Generate Ema Code" to generate EMA code for the current prototype network model. This is shown in Figure 14.3.

"Generate Ema Code" will generate EMA code. Save the code into a file named "chapter14-case1.em.c" in the model directory. Now you can modify this EMA file in order to create the network model.

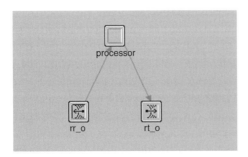

Figure 14.1   Node model

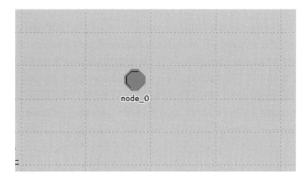

Figure 14.2   Network model

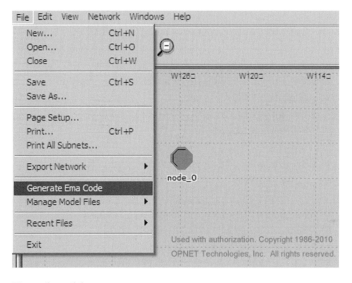

Figure 14.3   Network model

In "chapter14-case1.em.c" file, there are some routines for constructing an EMA program. These routines are general EMA functions starting with the prefix "Ema_". The details of EMA APIs can be found in "Programmers Reference" – "Model File Access API" – "External Model Access" section of OPNET documentation.

Before you can modify this EMA file, we will explain the existing EMA code. In the "chapter14-case1.em.c" file, the EMA code can be separated into five sections. The first section is the line with Ema_Init() function, which is used in every EMA program to initialize the EMA package. After "Ema_Init()" is invoked, other EMA functions can be used. The second section is the line with the "Ema_Model_Create()" function, which creates an empty model to be completed by adding other dependent objects to it. In this case, the model is a network model. The third section contains "Ema_Object_Create()" invocations. The "Ema_Object_Create()" function is used to create an instance of the model, i.e., an object. In this case, subnet-related objects, node object, and ISO elevation map color objects are created. The fourth section contains "Ema_Model_Attr_Set()" invocations. The "Ema_Model_Attr_Set()" function is used to set attributes for an object. These attributes are the same as you can see from the model instance's "Attributes" dialog via the OPNET graphic interface. Since all objects created here are instances of the network model, all invocations related to these objects should reference "model_id" returned from "Ema_Model_Create()". The fifth section is the line with "Ema_Model_Write()" invocation, which will finally write the completed EMA model into a model file within the primary model directory. The EMA-created model file can be loaded in OPNET

```
// Section 1
Ema_Init (EMAC_MODE_ERR_PRINT | EMAC_MODE_REL_60 , \
    argc, argv);

// Section 2
model_id = Ema_Model_Create (MOD_NETWORK);

// Section 3
obj [0] = Ema_Object_Create (model_id , \
    OBJ_NT_SUBNET_FIX);
obj [1] = Ema_Object_Create (model_id , \
    OBJ_NT_SUBNET_VIEW);
...
obj [4] = Ema_Object_Create (model_id , \
    OBJ_NT_NODE_FIXED);
...
obj [8] = Ema_Object_Create (model_id , \
    OBJ_NT_ISO_ELEV_MAP_COLOR_SETTING);
obj [9] = Ema_Object_Create (model_id , \
    OBJ_NT_ISO_ELEV_MAP_COLOR_SETTING);
...

// Section 4
Ema_Object_Attr_Set (model_id , obj[5] , \
    "elevation", COMP_CONTENTS , (double) 1e+100, \
    "color", COMP_CONTENTS , 1090519039 , \
    EMAC_EOL);

Ema_Object_Attr_Set (model_id , obj[6] , \
    "elevation", COMP_CONTENTS , (double) 5000, \
    "color", COMP_CONTENTS , 1073741824 , \
    EMAC_EOL);
...

// Section 5
Ema_Model_Write (model_id , "chapter14-case1");
```

Figure 14.4   EMA code

```
// Section 3
obj [0] = Ema_Object_Create (model_id, \
    OBJ_NT_SUBNET_FIX);
obj [1] = Ema_Object_Create (model_id, \
    OBJ_NT_SUBNET_VIEW);
obj [2] = Ema_Object_Create (model_id, \
    OBJ_NT_SUBNET_FIX);
obj [3] = Ema_Object_Create (model_id, \
    OBJ_NT_SUBNET_VIEW);

obj [4] = Ema_Object_Create (model_id, \
    OBJ_NT_NODE_FIXED);

obj [5] = Ema_Object_Create (model_id, \
    OBJ_NT_ISO_ELEV_MAP_COLOR_SETTING);
obj [6] = Ema_Object_Create (model_id, \
    OBJ_NT_ISO_ELEV_MAP_COLOR_SETTING);
obj [7] = Ema_Object_Create (model_id, \
    OBJ_NT_ISO_ELEV_MAP_COLOR_SETTING);
obj [8] = Ema_Object_Create (model_id, \
    OBJ_NT_ISO_ELEV_MAP_COLOR_SETTING);
obj [9] = Ema_Object_Create (model_id, \
    OBJ_NT_ISO_ELEV_MAP_COLOR_SETTING);
```

Figure 14.5　EMA code

```
// Declare variables
int base_index;
int node_count;

// Section 3
obj [0] = Ema_Object_Create (model_id, \
    OBJ_NT_SUBNET_FIX);
obj [1] = Ema_Object_Create (model_id, \
    OBJ_NT_SUBNET_VIEW);
obj [2] = Ema_Object_Create (model_id, \
    OBJ_NT_SUBNET_FIX);
obj [3] = Ema_Object_Create (model_id, \
    OBJ_NT_SUBNET_VIEW);
base_index = 4;
node_count = 100;
for(i = 0; i < node_count; ++i)
  obj [base_index + i] = Ema_Object_Create ( \
      model_id, OBJ_NT_NODE_FIXED);
for(i = 0; i < 5; ++i)
  obj [base_index + node_count + i] = \
      Ema_Object_Create (model_id, \
      OBJ_NT_ISO_ELEV_MAP_COLOR_SETTING);
```

Figure 14.6　EMA code

Modeler's graphic interface in the same way as any other model created directly via the modeler's graphic interface. The breakdown of these sections is shown in Figure 14.4.

To create the new network model specified at the beginning of this section, you only need to modify code in the third and fourth sections and leave other sections intact. Before modifying these sections, you need to include the header file "stdlib.h", since you'll use "rand()" function to generate random numbers. Change the size of the global array variable "obj" from current 10 to 109, since in current EMA code only one "ema_node1" instance is created while in the new network model there are 100 nodes. Then, you can modify the code in the third section. The existing code for the third section is shown in Figure 14.5.

We are only interested in the code that creates a node, as shown in Figure 14.5. The code for creating a node is replaced by a loop statement that will create 100 node instances of the "ema_node1" model. After 100 node objects are created, ISO elevation map color objects are also created via a loop statement. This modification is shown in Figure 14.6.

Next, you can modify the code in the fourth section to set the attributes for the 100 node objects created. These attributes include "altitude" attribute. There are many "Ema_Model_Attr_Set()" invocations in Section 4. However, you should only look for the one that sets the attributes for "node_0". This is shown in Figure 14.7.

```
// Section 4
Ema_Object_Attr_Set (model_id, obj [4],
    "name", COMP_CONTENTS, node_name,
    "name", COMP_USER_INTENDED, EMAC_ENABLED,
    "model", COMP_CONTENTS, "ema_node1",
    "model", COMP_USER_INTENDED, EMAC_ENABLED,
    "x position", COMP_CONTENTS, (double) -125.24,
    "x position", COMP_USER_INTENDED, EMAC_ENABLED,
    "y position", COMP_CONTENTS, (double) 6.2496,
    "y position", COMP_USER_INTENDED, EMAC_ENABLED,
    "doc file", COMP_CONTENTS, "",
    "doc file", COMP_INTENDED, EMAC_DISABLED,
    "doc file", COMP_USER_INTENDED, EMAC_ENABLED,
    "subnet", COMP_CONTENTS, obj [2],
    "alias", COMP_INTENDED, EMAC_DISABLED,
    "tooltip", COMP_CONTENTS, "",
    "tooltip", COMP_INTENDED, EMAC_DISABLED,
    "tooltip", COMP_USER_INTENDED, EMAC_ENABLED,
    EMAC_EOL);

Ema_Object_Attr_Set (model_id, obj [4],
    "ui status", COMP_CONTENTS, 0,
    "view positions", COMP_INTENDED, EMAC_DISABLED,
    EMAC_EOL);
```

Figure 14.7   EMA code

```
// Declare variables
char node_name[256];
int direction_x, direction_y;

// Section 4
srand(time(NULL));
for(i = 0; i < node_count; ++i)
{
  sprintf(node_name, "node_%d", i);
  rand()%2 == 0 ? \
      (direction_x = -1) : (direction_x = 1);
  rand()%2 == 0 ? \
      (direction_y = -1) : (direction_y = 1);

  Ema_Object_Attr_Set (model_id, obj [base_index + i],
      "name", COMP_CONTENTS, node_name,
      "name", COMP_USER_INTENDED, EMAC_ENABLED,
      "model", COMP_CONTENTS, "ema_node1",
      "model", COMP_USER_INTENDED, EMAC_ENABLED,

      "x position", COMP_CONTENTS, \
          (double) (-126 + direction_x*rand()%10),
      "x position", COMP_USER_INTENDED, EMAC_ENABLED,
      "y position", COMP_CONTENTS, \
          (double) (-6 + direction_y*rand()%10),
      "y position", COMP_USER_INTENDED, EMAC_ENABLED,
      "altitude", COMP_CONTENTS, (double) (rand()%1000),
      "altitude", COMP_USER_INTENDED, EMAC_ENABLED,

      "doc file", COMP_CONTENTS, "",
      "doc file", COMP_INTENDED, EMAC_DISABLED,
      "doc file" COMP_USER_INTENDED, EMAC_ENABLED,
      "subnet", COMP_CONTENTS, obj [2],
      "alias", COMP_INTENDED, EMAC_DISABLED,
      "tooltip", COMP_CONTENTS, "",
      "tooltip", COMP_INTENDED, EMAC_DISABLED,
      "tooltip", COMP_USER_INTENDED, EMAC_ENABLED,
      EMAC_EOL);

  Ema_Object_Attr_Set (model_id, obj [base_index + i],
      "ui status", COMP_CONTENTS, 0,
      "view positions", COMP_INTENDED, EMAC_DISABLED,
      EMAC_EOL);
}
```

Figure 14.8   EMA code

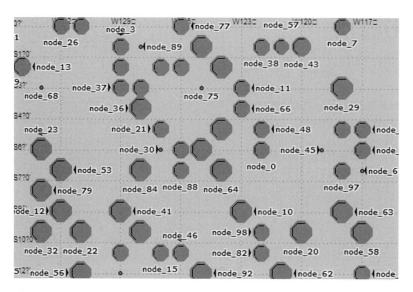

Figure 14.9   Network model

You should also modify this code in response to the modification made in Section 3, i.e., use a loop to set attributes for all 100 nodes created in Section 3. Furthermore, you can make random x and y position attributes for these nodes so that they will be randomly positioned on the network. For "altitude" attributes, they are random numbers between 0 and 1000. This modification is shown in Figure 14.8.

At this point, all EMA code modifications have been made. Next, you can compile and link the EMA program into an executable. Before building an EMA executable, you should make sure your "op_mkema" command's path is set in the PATH environment variable. Then, you can type the following command in console within your model's project path (assume the EMA source file is chapter14-case1.em.c):

$ op_mkema -m chapter14-case1

This command will build an EMA executable file in the project folder. This executable file is named "chapter14-case1.dev32.i0.em.x". Now, you can run this EMA executable to create model files:

$ chapter14-case1.dev32.i0.em.x

After running this executable, the model files will be generated in the project folder. Now you can open the project named "chapter14" from OPNET Modeler's graphic interface to load the EMA created network model. The new network model is shown in Figure 14.9.

These nodes are randomly positioned in the network. You can also check the "altitude" attributes of these nodes to see if they have different values randomly distributed between 0 and 1000.

15 OPNET co-simulation with third-party programs

In this chapter, the co-simulation capability of OPNET Modeler is introduced. The co-simulation interface allows OPNET Modeler to interact with external systems during simulation. These external systems can be software programs, hardware devices, or simply humans. Co-simulation is often used in situations where some third-party programs need to be used to process or analyze the intermediate simulation results, or some real-world data needs to be collected and fed back to OPNET Modeler during simulation.

15.1 Co-simulation with external programs

15.1.1 Introduction

OPNET Modeler provides a mechanism to support live interaction with an external system during simulation. This co-simulation mechanism involves the following concepts:

- **External System Definition (ESD) model**. The ESD model defines a set of interfaces that allow process models in OPNET modeler to communicate with external programs. These interfaces can be read or written by both OPNET process models and external programs.
- **Esys (External System) module**. The Esys module is a node domain module that can be placed into a node domain and processed as are other node domain modules, like processor module and queue module. However, the Esys module has extra features: it supports process models with external system communication capability and supports ESD models. The Esys module allows the user's simulation model to talk to external systems.
- **Simulator description file**. The simulation description file is a plain-text file containing statements that specify how to build co-simulation. This text file should be placed in the OPNET model directory and has the filename extension ".sd".
- **Esys API package**. The Esys API package contains functions that can read and write interfaces' values defined in the ESD model from process models during co-simulation.
- **External Simulation Access (ESA) API package**. The ESA API package contains functions that can read and write interfaces' values defined in the ESD model from external code during co-simulation. The external code refers to the code of an external system or a link to another external program. This package also contains functions that

are able to control the simulation flow process, read/write text, and issue debugging commands from external code.

- **External system**. External system represents any external program, device, or code that is in co-simulation with OPNET Modeler.

15.1.2   Co-simulation with an external C program

In this section, a case study is demonstrated to show how to write a co-simulation model that can communicate with an external C program in a bidirectional way.

15.1.2.1   Creating a simulator description file

A simulator description file defines statements that tell OPNET Modeler how to build co-simulation. The content of this file is shown in Figure 15.1. This description file is saved as "cosim_desc1.sd".

In the simulation description file, lines beginning with "#" are comment lines. "Platform" can be either "windows" or "linux_x86" depending on your operating system. "use_esa_main" can be either "yes" or "no". If it is "yes", the external program will be used in OPNET modeler as external dynamical libraries and entry point main() is declared in OPNET Modeler, i.e., the OPNET Modeler drives the co-simulation control flow. If it is "no", OPNET models are linked with the external program and entry point main() is declared in the external program, i.e., the external program drives the co-simulation control flow. "bind_obj" specifies the external program's object file that will be linked with OPNET models for co-simulation. "bitness" can be either "32bit" or "64bit", depending on your system's capability. "kernel" can be either "development" for debugging models or "optimized" for releasing models. "bind_lib" specifies the external static library to be used in co-simulation. "dll_lib" specifies the external dynamical library to be used in co-simulation. In this case study, only three statements are defined: "platform", "use_esa_main", and "bind_obj". "use_esa_main" is set to "no" to allow the external C program to act as main controller for co-simulation. "bind_obj" is set to "cosim_external_code1.obj", which is the compiled object file name of the external C program. The name of the external C program is "cosim_external_code1.c".

```
# Simulator Description
start_definition
    platform:       windows
    use_esa_main:   no
    bind_obj:       cosim_external_code1.obj
    #bind_lib:      xxx.lib
    #bitness:       32bit
    #kernel:        development
    #dll_lib:       xxx.dll
end_definition
```

Figure 15.1   Simulator description file

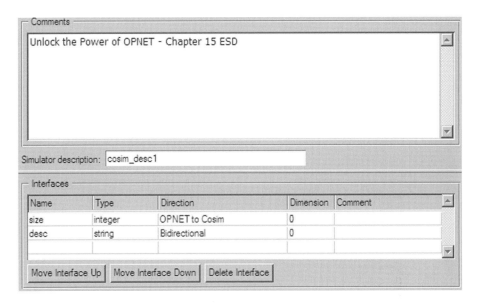

Figure 15.2   External system definition

15.1.2.2   Creating an external system definition (ESD) model

The ESD model defines all interfaces that are used to allow OPNET process models to communicate with an external program. To create a new ESD model, choose
"File" menu – "New..." – "External System Definition". For "Simulator description",
you should select the created simulation description filename "cosim_desc1". If it is not
shown in the list, you need to refresh model files by choosing "File" menu – "Manage
Model Files" – "Refresh Model Directories". For "Interfaces", we create two interfaces:
"size" as integer type and "desc" as string type. These types should be compatible C
language types in external C program, i.e., integer is int type and string is char[] type.
The direction refers to the way that OPNET models communicate with the external
program. If the direction is "OPNET to Cosim", it implies that OPNET models may
read/write interface and the external program may read interface only. If the direction is
"Bidirectional", it implies both OPNET models and the external program may read/write
interface. The configuration of this ESD model is shown in Figure 15.2. This ESD model
is saved as "cosim_esd1".

15.1.3   Creating simulation models

First, you can create an empty project and scenario with project name as "chapter15"
and scenario name as "case1". Next, you need to create the network domain topology
within the scenario "case1". The network has two nodes: one is traffic source node "src"
and another is traffic destination node "dest". This topology is shown in Figure 15.3.

The node domain logic for traffic source node is shown in Figure 15.4. The process
model of the "traffic_generator" module is "simple_source". The "Packet Interarrival

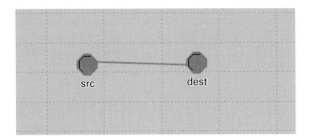

Figure 15.3   Network model

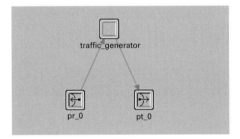

Figure 15.4   Node domain

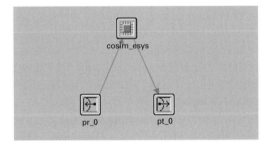

Figure 15.5   Node domain

Time" and "Packet Size" are "exponential (1.0)" and "exponential (1024)" respectively in this case.

The node domain logic for traffic destination node is shown in Figure 15.5. There is an external system module placed on the node domain. This external system module is responsible for communicating with the external program.

However, the actual communication code is implemented in the process model of this external system module. A process model called "cosim_process1" is created for this purpose. In the attributes of this external system module, you should set its "process model" to "cosim_process1" and "esd model" to "cosim_esd1". The state transition diagram of the "cosim_process1" process model is shown in Figure 15.6. Packet arrival will trigger state transition from "idle" to "strm". In "strm" state, interface values will be changed and the change will be notified to the external program. The external program

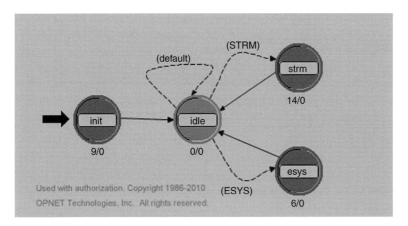

Figure 15.6 Process model

```
// State variables
Esys_Interface    inf_size;
Esys_Interface    inf_desc;
int total_pk_num;
char     desc[256];

// Temporary variables
Packet * pkptr;
```

Figure 15.7 Code in process model

may change an interface value as well. The change of interface value by the external program will trigger state transition from "idle" to "esys", where the changed interface value will be printed out.

This state transition diagram has two transition conditions. One is packet arrival interrupt. Another is external system interface interrupt. Packet arrival interrupt is scheduled when a packet is delivered to "cosim_esys" module. External system interface interrupt is scheduled when an interface value defined in the ESD model is changed and notified by the external program. There are two API packages related to co-simulation: Esys (External System) API package and External Simulation Access (ESA) API package. In OPNET process models, you should use functions in the Esys API package to read-/write interface value. In the external program, you should use functions in ESA API package to read/write interface value and/or control simulation flow. Therefore, in the "cosim_process1" process model, the Esys API package is used to read/write interface value. Figure 15.7 shows the declarations of state variables and temporary variables. Two Esys_Interface type variables are declared, corresponding to two interfaces defined in the ESD model.

```
total_pk_num = 0;

inf_size = op_id_from_name( \
    op_id_self(), OPC_OBJTYPE_ESINTERFACE, "size");

inf_desc = op_id_from_name( \
    op_id_self(), OPC_OBJTYPE_ESINTERFACE, "desc");
```

Figure 15.8   Code in process model

```
#include "stdlib.h"
#include "stdio.h"
#include "string.h"

#define STRM \
    (op_intrpt_type () == OPC_INTRPT_STRM)
#define ESYS \
    (op_intrpt_type () == OPC_INTRPT_ESYS_INTERFACE)
```

Figure 15.9   Code in process model

```
pkptr = op_pk_get(op_intrpt_strm ());

op_esys_interface_value_set(inf_size, \
    OPC_ESYS_NOTIFY_IMMEDIATELY, \
    op_pk_total_size_get(pkptr), 0);

sprintf(desc, \
    "No. of packets: %d - time: %f (process model)", \
    ++total_pk_num, op_sim_time());
op_esys_interface_value_set( \
    inf_desc, OPC_ESYS_NOTIFY_NEVER, &desc, 0);

op_pk_destroy (pkptr);
```

Figure 15.10   Codes in process model

Figure 15.8 shows the code in "init" state. In "init" state, the IDs of interfaces "size" and "desc" defined in ESD mdoel are obtained. With the interface IDs, you can access these interfaces via Esys APIs then.

Figure 15.9 shows the header block code. In header block, two state transition conditions are defined.

Figure 15.10 shows the code in "strm" state. State transition from "idle" to "strm" is triggered by the STRM condition. In "strm" state, "size" interface value is set to the

```
inf_desc = op_id_from_name ( \
    op_id_self(), OPC_OBJTYPE_ESINTERFACE , "desc");
op_esys_interface_value_get(inf_desc, (void *)&desc, 0);

printf("Process␣model\n%s\n\n\n", desc);
```

Figure 15.11 Code in process model

current received packet size, and the change of this interface value will be notified to the external program immediately. This interface value change notification will be posted to the external program via a callback function which will be invoked in the external program. For "desc" interface, its value is a formatted string that shows the total number of packets and current simulation time recorded in the process model. Although the value of "desc" interface is set, the value change notification is not posted to the external program, i.e., no callback function for "desc" interface is invoked in the external program after the interface value is changed.

Figure 15.11 shows the code in "esys" state. State transition from "idle" to "esys" is triggered by the ESYS condition. ESYS condition defines external system interface interrupt that is triggered by external program after "desc" interface value is changed. In "esys" state, "desc" interface value is printed out. The "desc" interface value can be changed in both the process model and the external program, since it is a "bidirectional" interface. However, the "size" interface is not printed in "esys" state. It is an "OPNET to Cosim" interface and therefore can be changed in the process model but not in an external program.

15.1.4 Creating an external C co-simulation controller program

In this case study, the external program is a C program which constructs a basic framework for interacting with OPNET modeler. However, co-simulation is not limited to C programs as the external program. This C program can be used to act as a co-simulation controller which controls the co-simulation flow and links the communication between OPNET Modeler's process model and another external program. An example of this case is demonstrated in Section 15.2, "Co-simulation with MATLAB." This C program is treated as an ordinary executable program, but functions in the External Simulation Access (ESA) API package are used in order to control the co-simulation flow and read-/write interfaces. Since "use_esa_main" is set to "no" in the simulation description file, this C program will act as main controller for co-simulation, and the co-simulation program entry function "main()" should be declared by this C program instead of by OPNET modeler. To use functions in the ESA package, the header file "esa.h" should be included in this C program. Figures 15.12, 15.13, and 15.15 show the code in main() function of the external C program.

In Figure 15.12, "EsaT_State_Handle object" is declared. This object is required in order to access ESA APIs. Two "EsaT_Interface" objects are declared. These objects are used to access the external system interfaces "size" and "desc" defined in the ESD model.

```
EsaT_State_Handle esa_handle;
int evt_num;
double ret_time;
EsaT_Interface *interfaces;
EsaT_Interface inf_size;
EsaT_Interface inf_desc;
int num;
int status;
int i = 0;
char desc[256];
```

Figure 15.12   Code in external program

```
// Initialize co-simulation and load ESA library
Esa_Main(argc, argv, ESAC_OPTS_NONE);
Esa_Init(argc, argv, ESAC_OPTS_NONE, &esa_handle);
Esa_Load(esa_handle, ESAC_OPTS_NONE);

// Obtain interfaces and register interface
// callback function
Esa_Interface_Group_Get(esa_handle, &interfaces, &num);
inf_size = interfaces[0];
inf_desc = interfaces[1];
Esa_Interface_Callback_Register( \
    esa_handle, &status, inf_size, \
    notification_callback, 0, 0);
```

Figure 15.13   Code in external program

In Figure 15.13, the first block of code initializes co-simulation and ESA library and loads the simulation network and associated files. The second block of code obtains the "size" and "desc" interfaces and registers a callback function for the "size" interface. This callback function will be invoked after the "size" interface value is changed in the OPNET process model "cosim_process1". The implementation of this callback function is shown in Figure 15.14. This callback function simply accumulates the value of "size" interface and prints it out.

Figure 15.15 shows the loop that controls the co-simulation process. The function "Esa_Execute_Until()" is used to run simulation until a specified simulation time is reached. Once the time is reached, "Esa_Execute_Until()" returns. After it returns, the "desc" interface value is obtained by the "Esa_Interface_Value_Get()" function and printed out. Then, a new value is assigned to the "desc" interface by the 'Esa_Interface_Value_Set()" function and a notification of this change is sent to the OPNET process model to schedule an interrupt by the setting ESAC_NOTIFY_IMMEDIATELY flag. This process will keep running until simulation

```
void notification_callback( \
    void *state, double time, void *value)
{
  printf( \
      "Total␣packet␣no.:␣%d,␣total␣size:␣%d␣bits\n", \
      ++total_pk_num, total_size += *(int *)value);

  return;
}
```

Figure 15.14 Code in external program

```
while(1)
{
  printf("External␣program\n");

  Esa_Execute_Until(esa_handle, &status, 10*++i, \
      ESAC_UNTIL_INCLUSIVE, &ret_time, &evt_num);

  if(status == ESAC_STATUS_TERMINATION)
  {
    printf("Simulation␣finished\n");
    break;
  }

  Esa_Interface_Value_Get(esa_handle, &status, \
      inf_desc, &desc);
  printf("%s\n\n\n", desc);

  sprintf(desc, \
      "No.␣of␣pkts:␣%d␣-␣time:␣%f␣(external␣program)", \
      total_pk_num, ret_time);
  Esa_Interface_Value_Set(esa_handle, &status, \
      inf_desc, ESAC_NOTIFY_IMMEDIATELY, &desc);
}
```

Figure 15.15 Codes in external program

time runs out or simulation is interrupted. In this case, the simulation interval for this loop is 10 seconds set by the "Esa_Execute_Until()" function. You can simply insert other external processing code within this loop and set corresponding interface values to feed back to the OPNET process model. In this way, the co-simulation process is completely controlled and the OPNET process model and external codes can easily communicate with each other via interfaces.

This external C program is saved as "cosim_external_code1.c" in your OPNET model's folder.

15.1.5   Running co-simulation

In this case, external codes are used for co-simulation, so we will bind OPNET simulation code, associated models and libraries, and external codes into an executable by using the op_mksim command. First, you should make sure OPNET Modeler's bin path is in the PATH variable in order to use OPNET modeler commands. If not, you can add the path to the PATH variable in the following way:

For Windows: path=%PATH%;[OPNET modeler win32 bin path]

For Linux: PATH=$PATH:[OPNET modeler unix bin path]

"[OPNET modeler win32 bin path]" should be replaced by the actual OPNET Modeler win32 bin path and "[OPNET modeler unix bin path]" should be replaced by the actual OPNET Modeler unix bin path. For Windows, it is better to use Visual Studio command line prompt tool to start a console session; otherwise, you need to add Visual Studio bin paths as well.

Next, you can compile the external C code into an object file. This is achieved by typing the following command in console under your OPNET project path:

For Visual C++ compiler:

$ CL /c cosim_external_code1.c /I"C:\Program Files\OPNET\[version]\ sys\include" /D_X86_
For gcc compiler:

$ gcc -c cosim_external_code1.c -I/usr/opnet /[version]/sys/include

"[version]" should be replaced by the actual version number of OPNET Modeler installed on your system.

This command will generate an object file "cosim_external_code1.obj" for Windows or "cosim_external_code1.o" for Linux within your project path. To bind this object file, simulation models, and libraries, type the following command in console:

$ op_mksim -net_name chapter15-case1 -c

The command option "-net_name" should be followed by the network model name. In this case, our project name is "chapter15" and the scenario name is "case1", so the network model name is "chapter15-case1". You can check OPNET documentation for more options of this command. If there are errors, you should resolve these errors before continuing. One possible error is failure of compilation of process models. For this error, you should check if there are compilation issues in your OPNET process models by hitting the "Compile Process Model" tool button in Process Editor. If there are no errors, this command will generate an executable called "chapter15-case1.dev32.i0.sim" in the chapter15.project sub-folder in the model directory. To run the co-simulation, type the following command line in console under your OPNET project path:

$ chapter15.project\chapter15-case1.dev32.i0.sim -duration 2000

The command line option "-duration" is to tell how long the simulation runs in seconds. The default simulation duration is 1000 seconds. The command options of the simulation executable are the same as op_runsim. For more options, refer to OPNET documentation for the op_runsim command.

The results of this co-simulation are: in OPNET process model, if a packet is received, the interfaces' values will be changed and notified to external code; external code will print out the value if an interface change notification is received in the form of a callback function; in external code, for every 10 seconds, an interface value will be modified and a notification will be sent to OPNET process model; OPNET process model will print out the value if an interface change notification is received in the form of a process interrupt. This case demonstrates how to make an OPNET process model to talk to external code to achieve co-simulation tasks.

15.1.6 Co-simulation with other systems

With this external C controller program, you can actually allow OPNET Modeler to interact with any external system that has basic inter-process communication capability. For example, if you want Modeler to communicate with a hardware device, you can write code in this controller program to talk to the software layer of the device via a socket or shared memory object. If you want OPNET Modeler to communicate with other open-source programs, you can invoke the public methods of those programs directly within this external C program. Furthermore, some programs provide direct C support, like MATLAB; for these programs, you can invoke the functions of these programs directly in this external C interface as well. To interact with humans, you can simply write code in this external C controller program to support command line read and write operations via scanf(), printf(), etc. If you want to write a large external program and prefer other object-oriented programming languages, you can write the program in your preferred language and build it into a library with C interfaces which can be addressed in this controller C program.

15.2 Co-simulation with MATLAB

Co-simulation with MATLAB is frequently used in OPNET modeling. For example, you are running a network to model a routing protocol in OPNET Modeler and the routing algorithm needs to be computed and analyzed at each intermediate node. However, the computation and analysis processes can easily be formulated in MATLAB, so you will want MATLAB to share the computation and analysis burden of the routing algorithm, and you will probably want MATLAB to plot the analysis results live during simulation. The MATLAB engine library contains routines that allow you to invoke MATLAB commands from your own programs. This engine supports C/C++ and Fortran programs directly (see www.mathworks.com/products/maths).

15.2.1　Setup of environment variables

Since the MATLAB engine library will be used in our co-simulation program, the relevant library paths should be added to the environment variable; otherwise, the co-simulation executable will not be able to find some required dynamical libraries during runtime. On Windows, they are DLL files. These DLL files include some OPNET dynamical libraries and MATLAB dynamical libraries. To add these paths, check "My Computer" – "Properties" – "Advanced" tab – "Environment Variables" button and add the "PATH" variable to the user variables. The value of the "PATH" variable should include the following two paths:

"C:\Program Files\OPNET\[version]\sys\pc_intel_win32\bin"

"C:\[MATLAB install path]\bin\win32"

Note: "[version]" should be replaced by the actual version of OPNET Modeler installed on your system and "[MATLAB install path]" should be replaced by the MATLAB installation root path.

Q15.1　Why do I get "libeng.dll" not found error when I run co-simulation executable?

This is because you did not set the paths of the necessary dynamical libraries to the "PATH" environment variable. If the co-simulation program uses the MATLAB engine library, the co-simulation executable will look for the dependent dynamical libraries at runtime. If it cannot find them in default paths, it will check the "PATH" environment variable for more search places.

15.2.2　Modifying OPNET models and external code

To prepare for co-simulation with MATLAB, you can modify the models and code made in Section 15.1. First, you need to modify the simulator description file so that the co-simulation program can be built with the MATLAB engine library support. You should make a copy of "cosim_desc1.sd" file and name it "cosim_desc2.sd". Modify "cosim_desc2.sd" so that it is the same as in Figure 15.16.

There are three statements different from "cosim_desc1.sd". For "bind_obj", the value is changed to "cosim_external_code2.obj". Furthermore, two static libraries have been added for build. The library "libeng.lib" is used for MATLAB engine routines and "libmx.lib" is for manipulating MATLAB types.

Next, in OPNET Modeler, open the external system definition (ESD) model
"cosim_esd1", then choose "cosim_desc2" in the ESD model and save it. This is shown
in Figure 15.17.

Q15.2 Why do I get "unresolved external symbol" link error when I use op_mksim
command to build a co-simulation executable?

This is because you did not include the necessary static libraries in the .sd file. You
should first investigate which third-party libraries are used, then include dependent

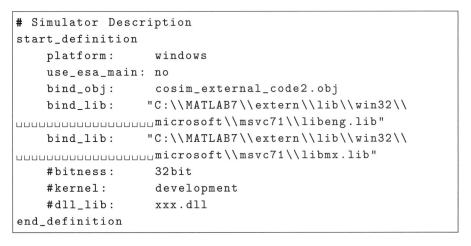

```
# Simulator Description
start_definition
    platform:       windows
    use_esa_main:   no
    bind_obj:       cosim_external_code2.obj
    bind_lib:       "C:\\MATLAB7\\extern\\lib\\win32\\
                    microsoft\\msvc71\\libeng.lib"
    bind_lib:       "C:\\MATLAB7\\extern\\lib\\win32\\
                    microsoft\\msvc71\\libmx.lib"
    #bitness:       32bit
    #kernel:        development
    #dll_lib:       xxx.dll
end_definition
```

Figure 15.16 Simulator description file

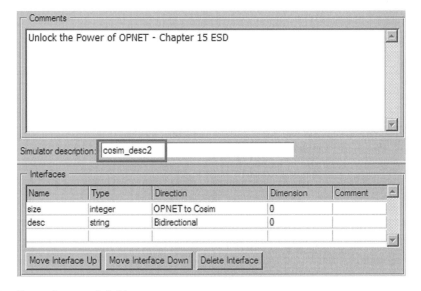

Figure 15.17 External system definition

```
EsaT_State_Handle esa_handle;
int evt_num;
double ret_time;
EsaT_Interface *interfaces;
EsaT_Interface inf_size;
EsaT_Interface inf_desc;
int num;
int status;
int i = 0;
char desc[256];
Engine *matlab_engine;
```

Figure 15.18   Code in external program

```
// Open MATLAB engine
if(!(matlab_engine = engOpen("\0")))
{
  printf("Fail to open MATLAB engine.\n");
  return 1;
}

// Initialize co-simulation and load ESA library
Esa_Main(argc, argv, ESAC_OPTS_NONE);
Esa_Init(argc, argv, ESAC_OPTS_NONE, &esa_handle);
Esa_Load(esa_handle, ESAC_OPTS_NONE);

// Obtain interfaces and register interface
// callback function
Esa_Interface_Group_Get(esa_handle, &interfaces, &num);
inf_size = interfaces[0];
inf_desc = interfaces[1];
Esa_Interface_Callback_Register( \
    esa_handle, &status, inf_size, \
    notification_callback, 0, 0);
```

Figure 15.19   Code in external program

libraries into the .sd file. For example, if you use MATLAB engine libraries in your co-simulation program, you should include "libeng.lib" and "libmx.lib" in the .sd file.

Next, you can modify the external C program to manipulate MATLAB. The first thing is to copy "cosim_external_code1.c" and name the copy "cosim_external_code2.c", which corresponds to the "cosim_external_code2.obj" set in the "cosim_desc2.sd" file. Then, you can modify "cosim_external_code2.c". You should make sure the "engine.h" header file is included in the "cosim_external_code2.c" file. "engine.h" declares symbols

```
while(1)
{
  printf("External␣program\n");

  Esa_Execute_Until(esa_handle, &status, 10*++i, \
      ESAC_UNTIL_INCLUSIVE, &ret_time, &evt_num);

  if(status == ESAC_STATUS_TERMINATION)
  {
    printf("Simulation␣finished\n");
    break;
  }

  // Interact with OPNET process model via interfaces
  Esa_Interface_Value_Get(esa_handle, &status, \
      inf_desc, &desc);
  printf("%s\n\n\n", desc);
  sprintf(desc, \
      "No.␣of␣pkt:␣%d␣-␣time:␣%f␣(external␣program)", \
      total_pk_num, ret_time);
  Esa_Interface_Value_Set( \
      esa_handle, &status, inf_desc, \
      ESAC_NOTIFY_IMMEDIATELY, &desc);

  // Handle MATLAB tasks
  handle_matlab_tasks(matlab_engine);
  array_index = 0;
}
```

Figure 15.20 Code in external program

of routines of the MATLAB engine library. In "main()" function, the MATLAB engine object should be declared. This is shown in Figure 15.18.

Figure 15.19 shows the code for initializing the MATLAB engine and co-simulation ESA library. The MATLAB engine routine "engOpen()" function is used to open the engine and return the pointer to the engine object if successful. This pointer will be used to access other engine routines.

Figure 15.20 shows the co-simulation loop where OPNET Modeler, external code, and MATLAB can interact with each other. Most of the operations are the same as in "cosim_external_code1.c", but a function "handle_matlab_tasks()" is added to the end of the loop. This function will handle all MATLAB-related tasks. In this case, "handle_matlab_tasks()" will dynamically plot a figure with received packet number versus packet size. The implementation of this function is shown in Figure 15.21. The "mxCreateNumericMatrix()" function is used to create a MATLAB-compatible matrix. The "engPutVariable()" function is used to assign a variable value from external code

```
void handle_matlab_tasks(Engine *matlab_engine)
{
  mxArray *x_pkt_num = NULL, *y_pkt_size = NULL;

  x_pkt_num = mxCreateNumericMatrix(1, ARRAY_LEN, \
      mxUINT32_CLASS, mxREAL);
  memcpy((void *)mxGetPr(x_pkt_num), \
      (void *)num_array, sizeof(num_array));
  engPutVariable(matlab_engine, \
      "x_pkt_num", x_pkt_num);

  y_pkt_size = mxCreateNumericMatrix(1, ARRAY_LEN, \
      mxUINT32_CLASS, mxREAL);
  memcpy((void *)mxGetPr(y_pkt_size), \
      (void *)size_array, sizeof(size_array));
  engPutVariable(matlab_engine, \
      "y_pkt_size", y_pkt_size);

  engEvalString(matlab_engine, \
      "plot(x_pkt_num,y_pkt_size);");
  engEvalString(matlab_engine, "hold on;");

  printf("Matlab operations finished.\n");
  mxDestroyArray(x_pkt_num);
}
```

Figure 15.21   Code in external program

to the MATLAB work space, so that you can use the variable the same way as in the MATLAB work space. The "engEvalString()" function allows you to invoke MATLAB commands and functions within external code provided that these MATLAB commands and functions are passed to "engEvalString()" in the form of strings. For details on MATLAB engine library routines, consult the MATLAB manual.

Finally, at the end of "main()" function, use the "engClose()" function to close the MATLAB engine. Now, OPNET Modeler can talk with MATLAB via an external C program which acts as a controller. The last thing is to run the co-simulation. Follow the steps below to build and run this OPNET modeler and MATLAB co-simulation project.

- In command line console, within OPNET model path, type the following command to compile the external C program:

   $ CL /c cosim_external_code2.c /I"C:\Program Files\OPNET\[version]\sys\ include"
        /I"C:\[MATLAB install path]\extern\include" /D_X86_

This will generate an object file called "cosim_external_code2.obj" in the current project folder.
- In command line console, type the following command to build the co-simulation executable:

$ op_mksim -net_name chapter15-case1 -c

This will generate an executable within the "chapter15.project" sub-folder.
- In command line console, type the following command to run the co-simulation:

$ chapter15.project\chapter15-case1.dev32.i0.sim -duration 5000

While the co-simulation session is running, MATLAB keeps plotting a figure by using the live data fed back from OPNET network simulation. At the same time, the OPNET process model and the external C program will print out exchanges from each other.

16 Model authoring and security

OPNET Modeler provides an authoring tool to allow authors to protect their models against unauthorized use. This chapter demonstrates how to use this authoring tool to protect and license user OPNET models. In order to follow this chapter, a reader should understand the basic concepts of OPNET models.

16.1 Introduction

In practice, you often need to publish models to allow other people to use them. However, sometimes you want them only to *use* these models but not to view the implementation details of these models, and/or to use these models for a limited time, like a demo version with limited access ability. To handle these issues, OPNET Modeler provides an authoring tool that allows you to encrypt your models to provide three levels of access: Use, Inspect, and Full. For "Use" access, users can only use the model in a simulation, but cannot inspect this model's structure or code via either the GUI editor or programming interface. For "Inspect" access, users can use this model in a simulation and inspect the model with certain programming interface procedures, but cannot view the structure of the model via the GUI editor. For "Full" access, users have full access to the model, i.e. they can use it, view it, and modify it in either the GUI editor or programming interface. By default, a model is not protected by the authoring tool, which therefore has "Full" access.

Model protection in OPNET Modeler is achieved via the "op_manfile" commands, which is a utility performing management operations related to the tracking and identification of model files, including file location, locks, security protection/registration, and header contents. However, in this chapter, we utilize only its ability to protect and register model files. For a full list of command preferences for the "op_manfile" program, type "op_manfile -help" in the OPNET command line console.

16.2 Protecting a model

First, a user needs to open the OPNET command line console. To protect a model with encryption, type the following command in the OPNET command line console:

```
In order to generate information which will be used
to encrypt the file and uniquely identify it, enter
at least sixteen characters at the following prompt;
the actual value is not important and need not be retained.
>
> │123456789qwertyuio│

Enter a descriptive name for the model or model suite.
Model (suite) name> │My First Model│

Enter one or more lines of additional information
(e.g. copyright, contact info, general comments, etc.);
enter a line containing only a period (.) to finish.
│.│
Success.
```

Figure 16.1   Command line

```
Available instances of model (mymodel) are listed below:

    0) Node Model
    1) Process Model (Portable Data)

Please enter model type index:
Index > │1│
```

Figure 16.2   Command line

> op_manfile -protect -m mymodel

where "mymodel" is the OPNET model to be protected. The OPNET model is saved as an .m file. For example, a process model will be saved as "mymodel.pr.m", a node model will be saved as "mymodel.nd.m", and a link model will be saved as "mymodel.lk.m". This command will prompt a user to enter a key for encrypting the model and ask the user for a descriptive name for the model, as shown in Figure 16.1.

It is possible that several models have the same name. For example, a process model and a node model can have the same name "mymodel", but their file names are different: "mymodel.pr.m" and "mymodel.nd.m" respectively. In this case, this command will prompt you to choose which model you want to protect, as shown in Figure 16.2.

The process model will then be protected. You can enter the following command to see authoring and security information for a model:

> op_manfile -header_print -m mymodel

```
Access to the model file will be enabled for all users.
Enter the desired access level (full, inspect, or use):

Access level > use
Enter the desired access expiration date
(enter "0" for indefinite access):

Expiration date (m/d/yyyy) > 1/1/2012

Success.
```

Figure 16.3   Command line

If you distribute this model to other client users, they will not be able to access it because of this protection. It is noted that the author always has full access, whether a model is protected or not. The protection is only against unauthorized client users.

16.3   Making a demo model

In some cases, you will want to publish your models. However, you want to give client users only limited access privilege and access time; for example, only allow them to use your models instead of viewing the details of your models or modifying them. To achieve this, you can enter the following command in OPNET command line console:

> op_manfile -register -world -m mymodel

This command will prompt you to choose an access privilege and expiration date for the model, as shown in Figure 16.3.

Now, you can distribute this model to your client users. They can only use this model in their simulation and they are unable to view the details of the model or modify it. Further, they are only able to use the model until January 1st 2012.

16.4   Licensing a model

In some other cases, you will want to distribute your models but only want your approved users to be able to access them by registering these models. This can be achieved via a licensing process which is a three-way handshake process. First, you should protect a model in the same way as in Section 16.2. Next, distribute this protected model to client users and ask them to send you back a transaction code. Client users can retrieve this transaction code by entering the following command in their OPNET command line console:

> op_manfile -register -m mymodel

```
Send the following transaction code to the model vendor:
Transaction code > │66AD.DA14.2AD1.B73A│
```

Figure 16.4   Command line

```
Enter the transaction code sent by the client:
Transaction code > │66AD.DA14.2AD1.B73A│

The client's group ID is 7356.
Enter the desired access level (full, inspect, or use):

Access level > │use│
Enter the desired access expiration date
Expiration date (m/d/y) > │0│

Send the following approval code to the client:
Approval code > │ DE56.0D3F.CB27.5D8D.68E0.A73B │
```

Figure 16.5   Command line

This command will produce a transaction code, as shown in Figure 16.4. Client users should let you know their transaction codes. In this case, the transaction code that the client user should send to you is 66AD.DA14.2AD1.B73A.

When you receive the transaction code sent from the client user, you she or he should run the same command on this model:

> op_manfile -register -m mymodel

This command will prompt you to enter the transaction code received from the client user and set an access level and expiration date for this model. Finally, it will generate an approval code which will be sent to the corresponding client user for registration. This is shown in Figure 16.5.

After the client user receives the corresponding approval code from you, she or he should enter the following "op_manfile" command again on this model for registration:

> op_manfile -register -m mymodel

This command will prompt the client user to enter the approval code to register this model. After registration completes, the client user can access the model according to the access level set by you; meanwhile, other client users are unable to access the model before having their own approval codes for it. In this way, you can license your models to approved client users only.

References

Bing B. *Wireless Local Area Networks: The New Wireless Revolution*. Wiley, 2002.

Hayes J. F. *Modeling and Analysis of Telecommunications Networks*. Wiley-Interscience, 2004.

Kleinrock L. *Queueing Systems*. Volume II C. Computer Applications. Wiley Interscience, 1976.

Leemis L., Park S. *Discrete Event Simulation: A First Course*. Pearson Prentice-Hall, 2006.

Park K., Willinger W. *Self-Similar Network Traffic and Performance Evaluation*. Wiley-Interscience, 2000.

Robinson S. *Simulation - The Practice of Model Development and Use*. Wiley, 2004.

Ryu B. A tutorial on fractal traffic generators in OPNET for internet simulations. OPNETWORK 2000, Washington D.C., August 2000.

Ryu, B. Lowen S. Point process models for self-similar network traffic with applications. *Stochastic Models*, 1998.

Sohraby K., Minoli D., Znati T. *Wireless Sensor Networks: Technology, Protocols, and Applications*. Wiley-Interscience, 2007.

Useful internet resources

www.boost.org/

www.gnu.org/software/gdb/

www.isi.edu/nsnam/ns/

www.mathworks.com/products/matlab/

www.microsoft.com/

www.omnetpp.org/

www.opnet.com

www.techsmith.com/snagit/

www.video2down.com/

Index